Journey
To
Reason

Walking Away from Young Earth
Creationism and Religious
Fundamentalism

Independently published by Mark Aaron Alsip.

Web: www.alsip.net
Email: publishing@alsip.net
Twitter/X: @markaaronky
Instagram: www.instagram.com/mark_alsip
Facebook: www.facebook.com/MarkAlsipAuthor
TikTok: www.tiktok.com/@mark_alsip

Dedication

To all those who have undergone religious indoctrination, especially the children and young adults. Hope, and help, are out there. Keep searching for reason. Never give up.

Contents

Part One

In the Beginning

"There is a cult of ignorance in the United States, and there always has been. The strain of anti-intellectualism has been a constant thread winding its way through our political and cultural life, nurtured by the false notion that democracy means that 'my ignorance is just as good as your knowledge.'"

— Isaac Asimov

Rainbows

I've never understood why some people don't find science classes exciting. One of my earliest memories was the day a grade-school teacher brought a shiny new object to class. He called it a "prism."

He walked over to a window and held a little glass object in a beam of sunlight and... magic! A rainbow appeared on the classroom wall. My jaw dropped. How could this be? My Sunday school teacher told me rainbows only came from God.

At this early level of school, it was necessary for teachers to explain things in terms a child could understand. We didn't discuss wavelengths or refraction. What I learned that day was that the white light we took for granted was actually made up of many components of different colors. Our eyes saw them all combined... as white light. But a prism could separate those colors and show us what really made up a beam of light.

I was excited. I wanted my own prism.

Living in a small Kentucky town, long before the Internet and Amazon, this wasn't exactly something you could pop over to the local store and purchase, or buy online. My parents weren't sure where to get one, but my father pointed out that there was an easy way to make a rainbow.

Unfortunately, this meant that I had to learn how to wash cars.

Dad would show me how to create a rainbow, but since we'd have the water hose out anyway and the car was dirty, why not kill two birds with one stone? Along with a science lesson, I'd learn the importance of doing household chores.

My father pulled the family station wagon up to the water hose and handed me a bucket of soapy water and a sponge. Then he did something amazing: he pointed the water hose into the air and let loose. A rainbow appeared! Satisfied, he left me to my car washing duties.

So I was tricked into washing cars, but I was still hooked on my new power. I could create rainbows any time the sun was up. (Perhaps I shouldn't mention my attempt to create rainbows in the bathroom at night with a shower—Mom and Dad were not happy.)

I went back to my science teacher.

"Yes," he said, "water droplets act like tiny prisms. Each droplet separates light in the same way I showed you in class." I got bonus points in school for being so interested in the subject matter. A young scientist was born.

Things didn't go so well when I showed up at Sunday school that weekend and announced that it wasn't only God who could create rainbows.

My Sunday school teacher frowned. (This wouldn't be the first time.)

"Well," she declared, "God made rainbows possible. Before the Great Flood, rainbows just didn't happen. You're experiencing a gift that God gave you."

This was confusing, even to a small child. If light was created as we were taught in Genesis 1, how was it that it didn't already contain all the components that I could separate with a prism or a simple water hose? Did the nature of light change at some point? Were raindrops once shaped differently?

(Later in life, in college physics and astronomy, I would gain the knowledge to undeniably show that rainbows existed before the flood.)

As usually happens in science, one answer leads to a plethora of new questions.

I went back to my science teacher. He seemed uncomfortable with some of my questions. I know now it's because I mentioned God. I was putting him on the spot.

When I went to grade school, classes began with daily Bible readings. Existence of God was ingrained in the culture. I feel bad for teachers who are prohibited from giving direct answers to questions. The list of prohibited topics for teachers continues to grow as I write these words. In a fundamentalist environment, you simply could not allow yourself to be seen as questioning God, or even public figures (politicians, religious leaders) who claimed to speak for our creator.

I could not have realized at the time that this was ominous foreshadowing for the future in which I now live. In 2024, laws are actually being passed to prevent teachers from discussing certain topics in schools. Educators are threatened with lawsuits and/or loss of their jobs for discussing topics that could be deemed "controversial." Simultaneously, fundamentalist outside forces are pushing ever harder—and often succeeding—in introducing religion into what is supposed to be secular curriculum.

My instructor was as honest as he could be, given the circumstances. "This is how light is made up, Mark. If you split it, you'll always get a rainbow. It's physics. That's how it works. That's how it's always worked." Ironically, this man was a devout Christian; a good teacher, and a good person. Like many of the non-fundamentalist Christians in my life, he saw no conflict in accepting science and religion. I would come to look back on this moment as an example of an early Christian role model. Constrained by society, he was still willing to speak the truth.

My Sunday school teacher wasn't pleased when I returned the next week and announced that she was wrong. I received a stern warning not to question the ways of God; Satan was tempting me. This rainbow nonsense was a clear attempt to sway me away from scripture, and if I didn't reject my scientific education, I was going to suffer the judgement of my creator. Read: Hell. End of subject.

This was the emergence of a pattern in my life. I would come up with questions and the Church couldn't give rational answers. I was not trying to cause trouble. I didn't mean to be argumentative. I believed in God. I believed in science. When there was a conflict between the two, it was inevitably resolved by my fundamentalist congregation threatening me with horrific outcomes if I didn't deny science. The problem was, in science I could perform my own experiments, repeat the results, and find truth. Wasn't my religion supposed to be a search for truth?

My parents would often find me standing in the yard, water hose spraying into the sky, my eyes locked onto the resulting rainbow in deep thought. "How could this have ever been any different?"

I planted a little garden in my head after this incident. A garden of doubt. The first seeds were planted underneath a beautiful (water-hose-induced) rainbow.

Mom and Dad continued to search for a prism. My experiments were running up quite the water bill.

The Altar

I was baptized as a Catholic soon after birth. This was the practice at the time. My mother was a good Polish Catholic from Milwaukee. My father was a Protestant from Kentucky. They met in Milwaukee while my father was in the army, part of the missile defense system designed to protect America from surprise Soviet nuclear attacks.

As a young child, I was puzzled by my parents' wedding photos. There was a long white wall in front of the church, waist-high, to prevent anyone from approaching the altar. I asked about this wall, and my mother explained it was because the Church said my father wasn't holy enough to approach the sacred altar. That privilege was reserved for "true Christians."

This was to be the first of many times in my life when I was told a parent, family member, or even myself, wasn't really a member of what I thought was the same religion.

As the story goes, Dad wasn't quite Christian enough for the Catholics. He believed in Jesus and went to church, but didn't recognize the pope as an authority, pray to saints, or believe in any of the myriad teachings of Catholicism. In the eyes of my mother's religion, my father may have been a good person, but he wasn't following the true religion. (Ironically, when my family later relocated to Kentucky and started attending Dad's church, his Protestant congregation would say the same thing about Catholics.)

To marry Mom, Dad had to attend regular classes, supervised by a priest, and agree that his children would be baptized and raised as Catholics. But he still wasn't in good enough standing with God to approach the altar and take marriage vows. So they were married in front of a wall.

Dad loved Mom enough to concede, and go along with the sleights that came with marrying her. Some in her family were opposed based on religious beliefs (and admittedly, Dad wasn't full-blooded Polish, which was another strike against him in the family... but that's another story for another time).

The Catholic church itself was fairly agreeable to a young child. I didn't understand Latin, which, by doctrine, was the language used in Mass at the time, but the after-party was wonderful. The basements of all the churches served beer and played bingo. Mom wouldn't let me drink beer (Uncle Bill slipped me a drink once), but I was allowed to play bingo. What wasn't to love?

I do have disturbing memories of Catholicism. When I asked about the photos of a man in a black suit sprinkling water over my head, it was explained to me that I was born a sinner. As I've admitted, my memories don't quite go back that far, but I don't recall ever doing anything wrong. How could I be responsible for sins I hadn't committed?

The statues and the saints were confusing. Grandma and Great-Grandma were always praying to statues, and telling me they were asking dead people (saints) to intercede with our Father so that their prayers would be heard. This was confusing, as I distinctly remembered a priest telling me I could talk to God or Jesus any time I wanted to.

Also disturbing to me were our nightly prayers. Kids today probably aren't familiar with the original version of this classic, as it's been reworded due to how frightening it was:

"Now I lay me down to sleep, I pray the Lord my soul to keep. If I should die before I wake, I pray the Lord my soul to take."

The last thing you want to do before sending a child to bed is to remind them they might die in their sleep. On many nights, my final conscious thoughts as I fell asleep didn't involve counting sheep—I was praying over and over that I'd wake up again in the morning.

Later, in my fundamentalist Protestant life, the bedtime prayer problem would be compounded. With children, the Protestants were much more Hell and Satan-heavy than the Catholics. My parents would often oblige my sisters and I by searching our closets and under our bed with flashlights to prove to us that there were no monsters hiding there. But then we'd have to pray, and I knew that the Devil was invisible. What if he was under my bed and we couldn't see him? What was that noise I just heard in my closet? I'd pull the sheets up over my head and tremble until I fell asleep.

I don't want to come out of the gate as sounding overtly anti-religious here. In both my Catholic church and my future Protestant one, I would learn valuable lessons about how to treat people; I was always surrounded by a group of people who I felt cared about me.

My problems with religion would grow primarily out of fundamentalist teachings, and concerns over vast rifts in what was supposed to be one, single, divinely revealed word of God. I ask the reader to keep in mind that it was religion itself that taught me to seek truth, and to speak out when I saw discrepancies. I'm fortunate to have many wonderful religious people in my life. If I happen to criticize a particular teaching that I've encountered, please keep in mind the unbendable, unwavering doctrine I grew up with. I do not want to accidentally paint with too broad a brush.

Regardless, my time as a Catholic was soon to come to an end. My parents decided we'd be leaving Milwaukee for Dad's home state of Kentucky. My religious education would continue, however, but it would now take place in what's known today as a Young Earth Creationist church.

As alluded to earlier regarding my Sunday school experiences, this is where the trouble really started.

Baptisma

One of the first things I learned when I started attending church in my new home of Kentucky was that I wasn't really a Christian. This was quite a shock. If you study modern theology, you'll discover some contradictions here. The Catholic Church says I was born with the original sin of Adam and Eve. My new evangelical church both rejected and accepted this premise... I wasn't exactly born with the sin, but all humans are born with the stain of sin, and only the brutal human sacrifice of Jesus could save us.

Regardless, my new church told me I wasn't a true Christian because... wait for it... I wasn't baptized by being fully immersed in water. The sprinkling I received as a child didn't count. Sorry Methodists (and others); according to fundamentalist evangelicals, you don't make the cut either.

The dispute, I learned later, was over the Greek word for baptism, "baptisma." Translated literally, it simply means cleansing in water. However, this was long before I started learning foreign languages and began to appreciate the nuances of translation.

As always, I had a helpful Sunday school teacher standing by to assist me.

Baptism, she explained, meant being *fully immersed* in water. Never mind the original Greek, of which she spoke not a word. Jesus was fully immersed in water when John baptized him in the

River Jordan. Baptism was a form of spiritual rebirth. How could you be reborn if you weren't first buried (in water?). This was one of my early introductions to the concept that it didn't matter what original Biblical texts said; what it really came down to was how an unordained church elder interpreted things.

So an early crack appeared in my religious education, one that would grow, and I will expand upon in future chapters: it wasn't just my own soul at risk. My mother, a devout Catholic, and absolutely the definition of the word "saint" by her loving actions, was suddenly not a real Christian. This was something that would haunt me and never go away. The ever-present Sunday school teacher explicitly told me: my mother was going to burn in Hell forever unless I could convince her to see the truth.

Thus began a recurring nightmare that would haunt me through most of my life. Me, sitting in heaven on a cloud, while my mom, the object of my unconditional love, a person who would without question give anything and everything she had to help another person, was going to burn eternally. In my dreams, I could hear her screaming from Hell. There was nothing I could do. She'd had her chance and had rejected it, so she deserved what was happening to her.

My parents couldn't understand why I would repeatedly wake up crying at night. I tried to explain as best a child could, but it didn't seem to register.

Sadly, this is not an isolated situation. Many modern-day evangelical churches and organizations hold to these teachings. One, but not the only example, is Kentucky-based Answers in Genesis, whose website boasts over forty million page views per year, and a claimed yearly visitor count of over one million to its religious theme parks.

With free admission for kids under ten.

That's a lot of frightened children.

Inquisitive Child

Adults didn't seem to know what to do with me as I grew up. By all accounts, I learned very rapidly. In some cases, I was praised for this. For example, my family was delighted by my love of reading. By third grade, I had advanced to reading adult books. In other cases, I was scorned—my first-grade teacher punished me for finishing our entire reader in one day and telling the whole class the story, including the ending where the Native Americans drove the buffalo over a cliff to end the hunt. Kids made fun of me for being a "know-it-all."

This dichotomy would repeat itself over and over in my struggles with the teachings of the church versus my natural inclination to question everything, and my early love for science. I noticed that religious teachers had the inclination to hold me back and/or "correct my thinking" when I presented problems and contradictions to them. On the other hand, important people in my life, some of my schoolteachers and definitely my parents, noticed my appetite for knowledge and fed it.

Some examples may help illustrate.

When I was eight years old, I discovered the formula for gunpowder in the encyclopedias my parents had purchased for their children. It all began as an interesting story on how the Chinese allegedly invented fireworks. Amazingly, the article listed all the ingredients. I was fascinated, and immediately set out to make my own firecrackers.

Charcoal was easy… plenty of it on the back porch next to the grill. Sulfur? Our local farm supply store kept it as nutrient blocks for cattle. It was easy to pick up shards lying on the floor when Dad wasn't looking. I didn't consider it theft—surely, they were going to throw the broken pieces away?

Saltpeter was a little more difficult. I found out that it was stocked at our local Rexall pharmacy, and stopped in one day after baseball practice to buy some. For some reason, the clerk was suspicious and refused to sell it to me. No matter, a good friend had some in a mineral and rock collection.

I didn't really understand purity, mixing instructions, or anything else I needed to know, but I did get a nice poof! I felt like a wizard. Don't tell Mom.

Things got a little more serious when I learned how to make poisonous chlorine gas from common household ingredients. My parents never fully understood some of the things I was doing, but Mom was suspicious when she saw me headed out the door with an armful of cleaning supplies. Fifteen minutes later, I did indeed have chlorine gas, and the burning lungs that went along with it. I did get busted this time. The choking gave me away. In retrospect, this experiment was not a good idea. But instead of punishing me, my parents did something amazing—something countless schoolteachers would repeat—but I would never encounter in church.

With my parents recognizing my interest in the subject, a brand-new chemistry set showed up under the tree one Christmas morning. Instead of discouraging my inquisitiveness, my mom and dad sought to channel it and encourage it. The lesson I took away from this is that there was no need to fear knowledge (though chlorine gas was forever prohibited from my list of allowed experiments!). Asking questions was a good thing. Just don't inhale.

This positive reinforcement was echoed by the schools I

attended. It would be impossible to thank the countless teachers who recognized someone in love with learning, and who encouraged that child. Except for the first-grade teacher who scolded me for being able to read well, my life was filled with teachers and mentors who saw a child with an insatiable thirst for knowledge, and who went out of their way to quench that thirst.

On the other hand, we had the church. I desperately wanted to please Jesus, God, and my congregation. Bible reading—*all* of the Bible—was a goal I set for myself, and with my reading skills and comprehension, I was quickly gaining the attention of my elders; enough so that I would eventually be tagged as a candidate for preaching in both our church and youth competitions across the Southeast. More on that to come.

For now, we had the problem that I was beginning to discover contradictions and errors in both scripture and in our preacher's Sunday sermons. My methods for applying strict logic and admitting/correcting my mistakes in chemistry did not go over well in a religious setting. Fundamentalist religion didn't seem to be interested in admitting or correcting errors.

Witness one of the countless incidents with my now frazzled Sunday school teachers. One Sunday, I was excitedly talking about dinosaurs (what kid didn't love dinosaurs?) when I made the mistake of explaining they lived hundreds of millions of years ago and died sixty-five million years ago.

"Oh no, Mark, the Earth is only 6,000 years old. It's impossible that dinosaurs lived that long ago."

"But we learned the age of the Earth in geology class in school. They have ways to understand the age of rock layers, and the fossils are found deep in layers that are very old."

Flustered teacher:

"Well, obviously, God put the dinosaur bones there to tempt us."

Equally flustered me: "But the book of James says that God tempts no man."

My goal of reading the Bible cover-to-cover was sometimes a blessing, sometimes a curse.

A severe point of contention throughout my Young Earth Creationist upbringing was anything and everything to do with fossils, and, by extension, Noah's flood. My church freely admitted that fossilized remains of countless creatures did, in fact, exist. They were there because of a global deluge that buried and wiped out all life on Earth. But, as I would point out in Sunday school, such a flood would have left a global debris layer of fossils made up of every creature on Earth. I came to this conclusion entirely on my own; I was never coached by an "evolutionist" teacher to say something like this.

I did ask though, and my science teachers assured me that no such layer existed anywhere on the planet. Fossils were separated, quite literally, by layers of time. I must point out again that my instructors were invariably Christians themselves, so it was only my fundamentalist church that was the odd man out. I wasn't trying to argue with my Sunday school teacher. I wasn't trying to disprove religion. I was trying to reconcile science and my knowledge of the Bible.

I should pause here to point out that my Young Earth Creationist church even contradicted what later pseudoscientific religious organizations, like Answers in Genesis, would claim: that Noah actually took dinosaurs with him on the ark. These were the days before the Internet, when misinformation didn't spread as quickly as it would in my future. Dinosaurs on the ark would have raised as many, if not more, questions in my mind than God planting bones to tempt us. Young earth creationism just hadn't "advanced" to a consensus on this point yet.

After this experience, I shrugged my shoulders and went home to play with my chemistry set. I always got repeatable results there, as opposed to my fundamentalist congregation, where it seemed like I was constantly presented with contradictions that ended in menacing statements that my soul was at risk if I didn't accept a certain viewpoint.

This was a very heavy burden to place upon the shoulders of a young child who was only motivated by a search for truth. My childhood was filled with the ever-present terror that I was displeasing God because I studied science.

Abraham and Isaac

Not all of my childhood conflicts were between religion and science. There were significant clashes between fundamentalism and basic morals as well. As I pursued my religious education, I couldn't help but notice that there was a lot of killing in the Bible. I'd already questioned my elders about the moral and ethical implications of all the innocent babies and children dying in Noah's flood. I was reassured that at least they got into Heaven.

I didn't question so much the deaths of all the firstborn children of Egypt. The Egyptians were the bad guys in that story; they had it coming, according to scripture. I learned these lessons in the naïve, innocent days of my youth when everything was black and white. As children, we played Cowboys and Indians every weekend, shooting down the "bad" Indians without a second thought. We went to movie theaters and cheered John Wayne for doing the same thing. Admittedly, our church never came right out and said that God had given the okay to kill Native Americans, but it was often pointed out that they weren't Christians, and thus an inferior race.

We *were* taught that it was God's will that the United States would become a Christian nation (a belief echoed by conservatives today). Anybody who got in the way of that manifest destiny… well…

Nor did we question the destruction of entire cities, including the killing of every man, woman, child, and even the livestock (1 Samuel 15:2-3). Once again, these were enemies of God who were dying, so it didn't bother anyone in my church. I prayed that at least the little kids got to go to Heaven, like the babies from the flood did, and pushed it to the back of my mind. As skeptical and inquisitive as I was as a youth, it was easy to compartmentalize and justify killing when it was evil people who were dying.

The wheels came off that bus, though, when we reached the story of Abraham and Isaac in church. Abraham was a faithful, godly man. God decided to test him by ordering him to sacrifice his beloved son, Isaac.

To my absolute horror, Abraham agreed.

From what I could glean from my Bible concerning Isaac, he was innocent and faithful. He'd done nothing wrong. Why was he going to be slaughtered? A parent killing their child? Would *my* parents sacrifice me if God told them to?

Abraham took his son away from their camp under the pretense of offering an animal sacrifice to God. (I never understood why our creator demanded blood sacrifices in the first place, but was told not to question this practice. This was God's world; he could do as he pleased, and we were here to obey.) Abraham commanded the servants that accompanied them to stop and wait. He proceeded along the path with his son. When they were alone, Abraham started stacking firewood for a burnt offering. Isaac asked where the sacrificial lamb was. His father's response was to tie him up and lay him out on the firewood. Abraham pulled a knife to slit his son's throat before burning him.

At the last minute, God intervened and told Abraham to stop. He'd passed a test. Abraham was good in God's sight because he was willing to sacrifice his own child. Conveniently, a wild ram showed up, entangled in nearby shrubbery, and it became the sacrifice instead.

This story frightened me, and added to the routine nightmares I was already having because the Church said I wasn't a true Christian yet. In addition to the threat of being burned alive for all eternity by God, I was now aware that he could tell faithful followers to kill not just enemies, but their own beloved family. And they would obey.

I was upset.

My religious teachers tried to comfort me by saying God never really intended for Isaac to die. This was just a test for Abraham. His son was never in any danger. All of the previous killings by God and/or in the name of God came bubbling to the forefront of my mind. This was too much to ignore. It didn't matter to me that Isaac was never intended to die.

What mattered is that someone was willing to kill just because God said so.

Our church fully accepted the morality here. I had great difficulty doing so. Even as a child, I felt that this was incorrect thinking. This was dangerous thinking.

Decades later, on a beautiful fall day, September 11, 2001, the world would receive a demonstration of just how dangerous this thought process really was.

A Lot of Doubt

The global flood, in which so many children must have died, and the story of Abraham and Isaac, are instances where dark shadows were cast across my religious well-being. The death of all the innocent firstborn children of Egypt played into this equation as well. I found it difficult to understand how easily my congregation accepted these events as morally correct.

I soon encountered another such troubling story, this time involving a man named Lot. Lot is described in the Bible as a righteous man. In one chapter of Genesis, he kindly offers food and lodging to men who are actually angels traveling in disguise. Lot happens to live in an evil city though, and soon a crowd arrives, surrounds Lot's house, and demands that he send the two travelers out so that the gang can rape them.

When we were told this tale in church, Lot's reply astonished me: he begs the crowd not to do something so wicked as raping two men, and offers instead to send out his own virgin daughters, with the promise that the crowd can do whatever they please with the young women.

To righteous Lot, the raping of men was wicked, but for his daughters, a separate set of morals seemed to be in play. How was it evil to do this to men, but not to women, and why was a righteous father offering up his own children to such horrors?

Thankfully, the angels stepped in and blinded the crowd so they couldn't see Lot's door anymore. But no commentary was offered on Lot's decision. Instead, being righteous, Lot is allowed to lead his wife and daughters out of the city as God destroys everyone else in it. Except, as we all know, Lot's wife takes a look backward and is turned into a pillar of salt for disobedience.

No punishment for Lot's attempted betrayal of his daughters, but a very bad ending for a woman who dared to look over her shoulder. The sins of Lot and his wife didn't seem equal to me; their punishments (or lack of) seemed disproportionate.

Lot's story became more confusing. He drank too much, had sex with the same daughters he offered to the crowd, and both girls became pregnant. The blame for this incident is placed squarely on the women, since they got their father drunk.

I found even deeper problems hidden in the story. This wasn't the first time we'd studied the collective destruction of an entire city, including its children. In Lot's story, God directly caused the annihilation of Sodom. In another book of the Bible, He would instruct a human army to destroy not only every living man, woman, and child, but also all of the livestock.

Our preacher obviously felt there was a moral lesson in Lot's story. Lot was the good guy because he obeyed God and didn't look over his shoulder. This man's goodness was confirmed by a jump to 2 Peter in the New Testament, where Peter confirms that Lot was a righteous man who was by his virtue delivered by his creator. Lot's suffering, in having to deal daily with the evil people of his city, was praised and admired. The preacher concluded we should all strive to be as righteous as Lot.

Not a single word about the problems inherent in offering your daughters up to be raped by a mob, or having sex with your own children.

In my fundamentalist upbringing, I was presented with example after example of questionable moral acts, but nobody else in my congregation seemed to see a problem. In their eyes, if it was in the Bible, then it was good. People who followed this book were good. These were facts that were not to be questioned. Even the Bible says not to question the Bible.

Tragically, acceptance of this type of thinking would play a part in one of the most devastating events of my life, which I will recount in the next chapter. Blind, unquestioning acceptance of good and evil based on the apparent holiness of a person would become a problem not only for one of the world's leading religions (I refer to the Catholic priest scandal), but for me personally.

Because I respect the feelings of my readers, I will offer a "trigger warning." I am about to discuss a topic that some may find distressing.

The Babysitter

I was sexually abused around the age of seven. This is the hardest chapter of this book to write. Seeing the words on the page brings back memories I've tried to keep buried for most of my life. It's not something I'd rather talk about, and in truth, I've had to question whether or not the story even has a place in the context of this book. I've concluded that it has.

My abuser was a neighbor, a teenage boy from a well-respected family. There were no warning signs. Judging by his girlfriends, he was very much into the opposite sex. He was a Christian, and attended church every Sunday with his family. He was well mannered, respectful, obedient to his parents… he ticked off all the boxes as a fine young man who could be trusted. In fact, as a kid, I looked up to him. As the only boy in a family with many sisters, I'd always wanted an older brother. He was very nice to me, and I looked up to him.

The abuse began when he was hired as my babysitter.

There I was one evening in my own home, with someone I trusted asking me to take my clothes off. Something told me this wasn't right. I knew from Sunday school and my parents that we weren't supposed to be seen naked. God didn't like that. Adam and Eve were ashamed, and so was I. But it's easy for an older, trusted authority figure to convince you to do something you don't want to do.

The babysitter followed a pattern I'd later learn was typical: this was to be a secret between me, him, and God. I should trust him and never, ever tell anyone else what happened. God would not be pleased with me if I lied.

Being seven years old and taught to respect God and my elders, I promised to be quiet about what happened.

There's no need to go into detail about what occurred next. Suffice to say it happened several more times—any time he had a chance to be alone with me.

I can't explain how or why I knew what was happening was wrong, and why I felt so unclean. No, "unclean" doesn't quite describe it; "filthy" would be a better adjective. I resolved to tell the babysitter this was going to end, and if he didn't like it, I was going to break my promise and tell my parents.

Some memories, no matter how old, seem to never go away. This was one of them. It was as if all the blood in the babysitter's face had drained to his feet.

He reminded me that I had also promised God I wouldn't say anything. This was another matter altogether. Even at that age, I knew not to lie to my creator. So, a compromise was reached.

"One more time," he said, and it will never happen again. He came up with the now-ridiculous scheme of writing his promise out on paper, to me and to God, letting me read it, and then ripping it to shreds and flushing it down the toilet to remove the evidence. The promise, he said, was still binding. I was asked to promise God one more time that I would never talk about any of this.

And so, I endured one final episode of humiliation, and then it was over.

Except, as any victim of abuse will most likely tell you, it is never really over. Even though my rational mind tells me I was in no way at fault, the scars from this incident would stay with me for the rest of my life. A seed of a feeling of insignificance and unholiness was planted in my mind, and no matter what, it would never go away.

It is not my intent here to place blame for what happened to me on religion. My abuser certainly did not do what he did out of any righteous or spiritual motivation or teachings. He did it because he was not a good person. But nobody suspected him precisely because outwardly, he proclaimed to be a good person. He was by all accounts a good Christian.

I also do not hold responsible my parents or any of the adults who cared for me. They could not possibly have known what was going on. Good parents do everything they can to protect their children, and for my entire life, I've felt I've had the most loving and caring parents possible. It's simply impossible for them to be there every moment of every day.

My purpose here is to point out that we, as a society, often look the other way when it is a person of "good moral standing" who's accused of vile acts. Preachers, priests, politicians, judges… even "fine young men" such as fraternity brothers… they seem to get a free pass when it comes to sexual abuse. Sadly, I've come to notice, it is the accusers who are shamed and scorned. This problem seems to be especially prevalent with women who come forward.

If it's possible to make it even worse, conservative evangelicals in particular have become quick to accuse others of crimes such as pedophilia without any evidence, simply because, for example, of a person's sexual orientation. As I write this book, drag queens are vilified as pedophiles in some religious circles. My abuser was not a drag queen. He wasn't a member of the much-maligned LGBTQ community. He was a straight fundamentalist Christian who dated girls and went to church every Sunday.

But, it's easier to make scapegoats of minorities. These people were considered to be sinners in the first place. They were beneath contempt, much in need of repentance and their creator's saving grace.

Vaccines

As a child, I took for granted the small circular scar on my upper left arm. I noticed my younger sisters had the same marking. It wasn't until my church suggested that this was a candidate for the "mark of the beast" as mentioned in the book of Revelation that I became concerned and asked my mother about it.

It turns out this was a minor scar left by a medicine known as a vaccine. It was to protect me against a terrible disease known as smallpox. Mom had had her own particular dread disease to deal with as a youngster. It was called polio. She showed me pictures of kids in metal tubes known as "iron lungs." I was horrified... the victims could never leave these machines; they were trapped there for the rest of their lives, laying on their backs, with the iron lung doing all the breathing for them.

Though my church was wary of vaccines, I quickly grew to appreciate them. We eventually came to study how a scientist, named Edward Jenner, had developed the first one well over a hundred years earlier. The idea was so clever, and so simple, that even a grade-schooler could understand it. We had something called an "immune system," and it could be trained in advance to recognize certain diseases and fight them off. Like many children, I wasn't a big fan of needles, but whenever my parents took me to the doctor for my shots, I held back the tears. I wasn't afraid. I thought of those poor kids in the iron lungs and would have begged for that shot if it hadn't been offered to me.

Not so in my church. Not only did some consider the mark on

my arm a sign of something sinister, there was a belief that we should rely on God to heal us… that we were somehow interfering with His design if we started tinkering with our bodies.

I would find this belief pervasive as I grew older. Science denial seemed to have strong roots in fundamentalist religion. If scientists said it was helpful, my church would say it was bad. This attitude, I believe, set the scene for what would happen in the future Covid pandemic. I don't think it was a coincidence that so many fundamentalist churches in the Bible Belt were so strongly opposed to masks and vaccines.

It seemed possible to get a religious exemption for almost anything, and people in my church often did. All you needed to do to get out of *anything*, was to claim it was against your religion.

With every passing year of my life, the public seemed to trust less and less in science, even as we made incredible achievements such as landing on the moon and eradicating highly contagious diseases such as measles.

Sadly, because of vaccine denial and religious exemptions, even measles would make a comeback.

Many decades in my future, a disease known as Covid would kill millions of people. A large segment of the population, apparently with very little science background, would fight tooth and nail against a medicine that could help them. At the risk of painting with too broad a brush, much of the opposition to vaccines and other measures seemed to come from the evangelical right. One of the most common complaints where I live, in the Bible Belt, was that people couldn't attend church because of the prohibitions on public gatherings.

Preachers were constantly in the news repeating exactly what I'd been told as a youth in church: we didn't need vaccines; *God would protect us.*

I promised in the description of Journey to Reason that many of the problems tearing at our society in 2024 could be traced back to fundamentalism. Vaccine denial, and the broader issue of being able to exempt yourself from nearly anything based on your claimed religious beliefs, are two prime examples.

Jonestown

The year was 1978; my twelfth trip around the sun. As was tradition, my family gathered around the television to watch Walter Cronkite and the evening news before the "fun" television programs came on. My little sisters didn't care much for newscasts, preferring to play with their toys instead. For reasons I can't explain, perhaps it was just my fascination with the incredibly different world outside our small Kentucky town, I was addicted. There was a lot I didn't understand, but I was blessed with very patient parents who didn't mind answering questions.

One night, a horror story unfolded on our TV screen. It started out with a U.S. congressman and his entourage being gunned down on an airstrip in Guyana, a place I'd have to look up on a map. The incident led to investigators storming the compound, Jonestown, where the shooters were known to have come from.

Jonestown was founded by Jim Jones, a charismatic leader who had firm roots in Christianity. However, as his history unfolded, it became apparent that he was known for a strange ideological mix that would later be referred to by some as "apostolic socialism."

Regardless of naming conventions I was still too young to understand, what was found by the first crews to arrive at Jonestown was a nightmare beyond imagination: almost 1,000 people dead from apparent suicide. The phrase "drink the Kool-Aid" was born. First reports were that all these people had lined up to voluntarily drink the cyanide-laced beverage, after first making sure their own children had finished their shares. It would come out that Jones' followers had sometimes practiced their suicide ritual, having been lined up and obediently drinking what they were told was poison—but really wasn't—just to test their faith.

A test of faith… I couldn't help but think of Abraham and Isaac.

I asked my parents how this could have possibly happened. We talked about "brainwashing," giving in to false teachings that could lead a person astray. The topic would come up again in the following weeks at church as the full scope of Jonestown became apparent. We were warned against false prophets. Preachers focused mainly on Jones' early Pentecostal leanings (our church was fundamentalist, but no offense intended, even *we* considered Pentecostals to be right-wing). Sermons concerning Jonestown inevitably referred to Matthew 24:24, where Jesus warned that false Christs and prophets would appear. Jim Jones was, to our church, a prophecy fulfilled.

This set my young mind to wondering. I knew that one day Jesus was going to return to Earth. If there were going to be false Christs and false prophets who could perform miracles, how was I ever going to know the difference? Not yet dissuaded by my church elders' propensity to brush me off when I asked tough questions, usually with "just pray on it" when they could no longer answer with logic or Bible verses, I broached the subject.

Predictably: "Just pray on it, Mark. You'll know the difference. Trust us. Pay attention to what you're hearing in church. When the time comes, you'll just know."

I couldn't help but wonder: when the time came, did the followers of Jim Jones "just know?" What was the difference between one charismatic leader claiming to be preaching the truth and another charismatic leader doing exactly the same thing?

"Well, for one thing, God would never command us to kill ourselves or each other," my church said.

But... Abraham and Isaac! My elders were contradicting themselves.

I went home and prayed on it.

I found no answers. I saw danger signs. What would happen to our country, indeed, to the world, if enough people fell for a charismatic leader who made claims backed by fundamentalist religion?

Thief

My home town was very small. Over the sixty years of my life, the population has remained around 10,000 people. They don't ever seem to have to do much work on the "Welcome to Corbin" sign. I think they use the same stencil for repainting. The population numbers never change.

I'm not sure there was an actual furniture store in Corbin when we first moved there in the late 1960s. When I went shopping with my parents, it seemed that every shop had a hodgepodge of goods. Sometimes you'd see furniture on sale alongside 2x4s, nails, and other construction supplies in a home supply store.

I keep mentioning furniture, and this particular home supply store, because this was the setting for the first theft I'd ever make in my life. I couldn't have been more than six or seven years old at the time.

While my dad was busy talking to his brother, who worked behind the store counter, about whether twelve-penny or sixteen-penny nails would be better suited for a construction task at home, I wandered into the small furniture section and noticed a small decoration (I think it was a two-inch-tall Christmas tree) sitting in the middle of an end table that was for sale.

I liked the tree. It was pretty. Christmas was coming.

I knew that stealing was wrong, but this didn't seem like *stealing*. The tree was so tiny. It couldn't have been worth much. Surely this didn't really matter. I put the tree in my pocket.

Dad finished his business and we returned to the car. Somewhere, on the road home, I took the tree out and began playing with it.

"Where did you get that, Mark?"

"Uhm, I found it in the store."

"You *found* it?"

"Well, yeah, kind of. It was just sitting there on a table and nobody was using it so…"

Dad hit the brakes, turned into the first available parking lot, and headed back to the store. He was angry. "This family doesn't steal. We pay for things. You're taking that back."

Uh oh.

We pulled into the parking lot. Dad stopped the car. I opened the door and looked at him, waiting for him to join me. He didn't budge.

"Well, go on."

"What?"

"Walk into that store, go up the counter, and tell them you stole the tree. Apologize. Make sure they know you're sorry. And promise you're never going to do it again."

"You're not going with me? I'm sorry. I promise I'll never…"

"No, don't tell me. Tell *them*. Now go."

This was embarrassing. My uncle worked at the store. If I didn't do this right, Dad was going to know about it.

Shaking with fear, I opened the door, entered, and walked up to the counter. My uncle and the others behind the counter looked down at me. From the look on my face, they knew something was wrong. I was on the verge of tears.

"I'm really sorry. This wasn't mine, and I took it. I stole it off the table over there. I didn't mean it. I'm really, really sorry." I handed the little tree back.

The men behind the counter smiled at me. "Thank you for being honest. You're going to turn out to be a fine young man."

That was it. That's all there was to it. I walked back out to the car and got in. I looked up at my dad. I was expecting to really get chewed out now.

"I'm really proud of you." That's all he said. He started the car, pulled out onto the street, and we went home. The incident was never spoken of again.

Some people claim we need a church to teach us our morals, to tell us what's right and what's wrong. I'm not sure I agree with that now.

I think what we really need are good parents.

I began to wonder where morals came from. Were they subjective or objective? Did God hand down everything we needed to know? My church certainly told me so. As I grew up, I would grow to question some of the moral teachings of my congregation, but not my mother and father.

Today, I remember a frightened child being shown right from

wrong in a compassionate way. A way that would stick with me forever.

In my heart that day, I began to realize the true source of my morals.

Genesis

As I mentioned previously, I began a cover-to-cover reading of the Bible at an early age, and would re-read it throughout my life. I was already being fed bits and pieces through Sunday school, sermons during church services, and an interesting mail-order Bible study course originating in Tennessee, a sister state here in the Bible belt. Partly because I felt I wasn't getting enough, partly just for the challenge of it, I decided to read the whole book, not just the chapters and verses I was being fed by others.

Ironically (I'll elaborate on this in a later chapter), when my church found out I was doing this, I was warned not to.

One of the most-asked questions of atheists today is, "How/why did you become an atheist?"

"I read the Bible" is the most common answer. Back in my evangelical Christian days, I would have most likely countered this response with "Well, you haven't read all of it. You're just taking part of it out of context." This was my mindset when I plunged into that book—I wanted to know every piece of it. I didn't want to come up lacking. I took very seriously the wisdom of 1 Peter 3:15, which I came across in mail-order Bible study: *always be prepared to give an answer to those who question.*

The problems started cropping up, literally, *in the beginning*: the book of Genesis.

In Genesis 1:3-5, God creates light and divides the light from the darkness. Night and day were created. But it took over ten more verses (verse 16) for Him to create the sun and the moon. How did we have night and day without a sun?

Then came Genesis 2. Something seemed... off. Flipping back to Genesis 1, I discovered that chapter 2 had the order of creation wrong. Or was it chapter 1 that was wrong? They certainly disagreed.

In Genesis 1, God creates the plants before he makes Adam. In Genesis 2, Adam is created before the plants. God does put seeds in the ground for plants, but they hadn't grown yet. There was no water.

Chapters 1 and 2 are very confusing on whether dry land or water came first. In Genesis 1, we get water, which is then divided by dry land. Chapter 2 has dry ground first, which is then watered by a mist.

God also creates Adam all alone in chapter 2, puts him in the garden of Eden after finally watering the seeds, and tells Adam not to eat from the tree of knowledge. Eve isn't created until later on. But in chapter 1, Adam and Eve are created—and receive their instructions—in the garden together.

All of Earth's living creatures are created before Adam and Eve in chapter 1. But in chapter 2, God first creates Adam, puts him in the garden, tells him to stay away from that one tree, creates all the animals, and then, almost as an afterthought, He finally creates Eve. Apparently, in chapter 2, the animals by themselves weren't enough to keep Adam, who was created first, company.

It made my head spin to read and re-read this text (and it still does). Here I was, just two chapters into my project, and I was already finding contradictions. At church, I was told that this was the divinely inspired work of God, and that every word was true.

But if every word was true…

I was a huge fan of Star Trek when I was a kid. There's a famous episode, *Liar's Paradox*, where Kirk and Harry Mudd cause a malicious android to self-destruct. The conversation goes something like this:

Mudd: "I lied."

Kirk: "Everything Harry tells you is a lie."

Mudd: "Now listen to this carefully. I am lying."

Robot: "You say you are lying, but if everything you say is a lie, then you are telling the truth, but you cannot tell the truth because anything you say is a lie. But if you lie, you tell the truth… Illogical! Illogical!"

The android's head is enveloped in a cloud of smoke as his circuits fry.

Reading Genesis 1 and 2, I appreciated how that android felt.

Please understand that at this point in my life I had not set out on a journey to debunk the Bible. My intentions were pure. I was not looking for contradictions. I fully believed what my church was teaching me. I just wanted more knowledge. Like the understanding of the workings of the prism that ignited my love of science, I wanted to dig deeper into the foundations of my faith and be able to *explain*. In science and history classes, I learned that rational thought was good. Why shouldn't this apply to church as well? To be able to explain both the visible spectrum of light and the spectrum of God's creation: that was my earnest goal.

My project was off to a rocky start.

The Telescope

Chemistry wasn't the only science subject that fascinated me as a child. I was in love with astronomy. At the time, our family lived in the country, with our house backing up to farmland. There was virtually no light pollution. The Milky Way was a beautiful hazy ribbon that stretched overhead. I was fascinated to learn that that milky whiteness was actually a host of stars that couldn't be individually resolved with the naked eye. So many stars! I spent many a night stretched out on a lawn chair in the backyard, fighting off mosquitoes, staring at the sky and just… absorbing it all.

It may sound like a contradiction now, but early on, the night sky had me firmly convinced there was an all-powerful creator who put everything in place and set it into motion. I still remember watching my first lunar eclipse, along with my cousin. Lunar eclipses take a long time to develop compared to solar eclipses. As we sat there for an hour watching the shadow of the Earth wash over the Moon, realizing I was watching the actual shadow of our planet crossing the surface of another heavenly body, I turned to my cousin and asked, "How could anyone see all this and not believe in God?"

Our local library had a nice selection of astronomy books, and even subscriptions to *Sky & Telescope* and *Astronomy* (when it was finally published… *S&T* started much earlier). My typical routine was to take a large paper shopping bag to the library once a week—yes, these were the days before plastic bags—and fill it up with books for the week. Initially, this caused some pushback from the head librarian in our small town, who refused to believe a child could read that many books, that fast. And I was bored with the children's section, so I was pulling adult books. I finally convinced the reluctant librarian by launching into long discourses on the material I'd checked out the week before, and she relented. I was allowed to take home as many books as I wanted.

Word to the wise: old paper shopping bags don't really hold heavy books any better than the modern plastic bags. Especially on wet, rainy days. Another story for another time.

The astronomy magazines I had to read in-house. They weren't available for checkout. It was exciting reading. There was speculation that there may be other planets out there orbiting other stars, and that technology might soon be able to detect them!

This led me into making another mistake. I mentioned this news in church during a discussion of how special the planet Earth was. You could have cut the scorn and derision with a knife. "God only made the nine planets; Earth was special among them; it says so in the Bible, Mark."

First of all, the Bible says nothing about the number of planets. It seemed like every statement that came out of the mouth of those who I was told were religious authorities knocked them down further in my esteem. If a child knew this, and they didn't, well… somebody's been reading their Bible, and somebody is just making stuff up.

And yes, I said nine planets. Pluto hadn't been demoted yet. I may lose some of my scientific-minded readers for saying this, but I'm still rooting for Pluto. Again, another discussion for another time. At least the church didn't threaten the Pluto demoters like they did the early astronomers like Galileo, when he announced discovering amazing new moons.

Anyway, reading about the planets, I started yearning to see them with my own eyes. Yes, I had learned that certain bright objects changing their positions in the night sky were planets, and I could identify them. And there were plenty of pictures in my books and magazines. But I wanted to *see* them.

I've mentioned how wonderful my parents were for recognizing their children's interests and encouraging them. Late one evening, after supper, Mom and Dad walked into the kitchen carrying a long wooden box. "This is for you, Mark. Open it up."

A telescope!

It was just a small refractor, but it meant the world to me. Even the box that held it was a thing of beauty. Nicely finished wood, a felt-lined cradle with locking mechanisms to hold the instrument in place, a solid hand-crafted wooden tripod instead of cheap, thin metal. I was in awe. I ran outside with it right away.

And, because it was dark, I ran right back inside for a flashlight so I could go back out and assemble it.

My first views of the moon were amazing. Again, this was a small telescope, but I was looking at the moon with my own eyes! I tried Mars next. Astronomers familiar with small refractors know I was disappointed. A dull red dot became a... small red dot. My scope wasn't powerful enough.

But oh! The rings of Saturn! Not quite up to the level of textbook pictures, but there they were! And Galileo's moons orbiting Jupiter. There they were! I went back night after night and noted they were changing positions. A man was put under house arrest for this? Why was the church always doing bad things to scientists? Later I'd learn about a brave man named Bruno who was burned alive by his church for daring to claim that the universe didn't orbit the Earth. I resolved to be more careful in discussing science at church. I was already known as a troublemaker for doing so.

Of course, I turned my telescope on individual stars, knowing full well that they were still just going to be pinpricks of light. The stars were just too far away to resolve detail with a telescope. But my fascination was in knowing, from my reading, how far away these stars were. The light entering my eye had left before the United States was founded. Before Jesus walked the Earth. Maybe before the dinosaurs lived. Thousands, tens of thousands, millions, or even billions of miles away. And it had taken that light all that time to reach me in my backyard of Kentucky!

As would happen so often, I got overexcited, forgot my resolution to remain quiet, and mentioned this in church.

Oops again.

"No, Mark, you don't understand. The stars you're seeing were created 6,000 years ago when God created the universe. It's all right there in the Bible."

"But we learned the speed of light in the 6th grade. That can't be right! I've got books at home that explain how we know the distance to these stars! That light has to have been traveling for…"

"Mark, are those books the Bible? Are you putting the word of man before the word of God?" It was once again made very clear to me what happened when you questioned holy scripture.

Frustrated, and more than a little terrified, I gave up. I'd read the Bible far more times than these people. Not a single word said that the Earth or the universe was 6,000 years old. (Yes, I know the flawed genealogical calculations that led to this belief. You've heard me say it… another story for another time.)

The moral of this part of my story? I started out with a deep appreciation of the heavens and had total faith in the religious claims for its beginnings. I ended up with solid, concrete reasons to further distrust the people who were claiming to speak for God.

Back in my garden of doubt, more seeds were planted.

Roe v. Wade

The year was 1973. I was well into my immersion into Christianity and Young Earth Creationism. The news story of the day was a Supreme Court Decision that ruled women could have abortions.

My church was very upset that the government now allowed the "murdering" of babies. Being fairly well versed in the Old Testament at this point, I once again got into trouble in Sunday School. As always, this was unintentional. I was just asking questions.

The topic of the day was Noah's Flood. God had wiped out all but eight people on Earth, by drowning. This was very disturbing to me.

I am hopeless as a swimmer. My father was never cruel enough to try the old Southern tradition of throwing his kids over the side of the boat (yes, believe it or not, people do that), but he did try patiently to teach me to swim. We started around age four and continued until I nearly drowned in a lake as a teenager. I still remember the water going down my throat. No more swimming for me.

The bottom line was, I came to understand how horrible drowning (or near-drowning) could be. As later reading would confirm, it's one of the most horrifying ways to die. So, based entirely on my personal experience, I asked our Sunday school teacher why God would intentionally drown everyone.

"Oh. Because humanity was evil and turned against God."

"But," I replied, "even the little babies and children? Surely, they didn't do wrong? They didn't even understand what 'wrong' was."

My teacher got that perplexed look. The one that I came to know so well.

A long pause. Then:

"It wasn't like that, Mark; God understood the children weren't evil, so he took them all to Heaven." (But he drowned them first! I didn't go there. Yet.)

This is when I said the wrong thing. To be clear, I was only ten years old, so I didn't fully appreciate the weight of my words.

"So... all the babies that are being aborted now... are they automatically going to Heaven?"

The faces of all my Sunday school classmates went pale. Please understand, I was not trying to trap my teacher. I was not trying to ridicule God's word. I was a ten-year-old child who both read the Bible and watched the news. I was doing my best to understand a complex subject. I was simply a young, inquisitive kid, trying to wrap my brain around the moral implications of killing young children. I was worried about the babies in Noah's flood. I was worried about the babies our church said were now being killed.

My teacher's face... purple-faced consternation would be an accurate description. I'll never forget her stuttering, though I didn't understand a single word that came out of her mouth. Later in life, I learned the signs of the onset of a stroke. Had 911 existed at this time, and if cell phones would have been invented, I would have immediately called for help.

After the long pause that I grew to dread: "You don't understand, Mark. It's God's decision as to who lives and dies. With abortion, humans are making that decision. That's murder."

"But God says 'Thou shalt not kill', right?"

"Well, in this case, it was God who made the decision, so that rule doesn't apply."

Unfortunately, I pressed on. "So, God decided the babies and children would die, and they went to heaven. And the aborted babies are going to heaven. The news says the aborted children (I didn't understand 'fetuses' at the time) don't feel anything. I almost drowned once, and it was horrible. So why did God make all the children in the flood suffer through drowning when the aborted kids didn't have to suffer the same way? Didn't they feel it when they drowned? Why didn't God just take them up into Heaven?"

Once again, the merciful buzzer sounded, announcing the end of pre-service classes and the beginning of the main church services. Sunday school was dismissed.

I was aware of my teacher's eyes boring into my back as I walked out of the classroom.

I don't think that adults comprehend just how much children understand, and that at least some are capable of processing complex thoughts. They don't consider how their words affect these kids. My only concern at this momentous time in 1973 was the well-being of babies. I didn't dispute what the church told me about abortion. I was trying to reconcile two events that seemed identical in their outcome.

I was simultaneously told that the death of infants was both horrible, and yet sometimes part of God's will. As so often happened in my fundamentalist upbringing, another heavy moral burden was placed on my already sagging shoulders.

The Savannah

The moral problems with Noah's flood bothered me deeply. Drowning innocents didn't sit well with the morals I was being taught every Sunday. I took some comfort from the fact that my teacher said they went straight to Heaven. It seemed like a cruel way to go, but at least they got to be with God. I felt bad for all the new orphans though—according to the story, their parents were wicked in God's sight, so clearly, they didn't make the trip with their children. Could you be happy in Heaven knowing your loved ones weren't there? This question would often rise up and hit me from different directions throughout my Christian years.

But another of the many problems with the Great Flood was apparent to me, even as a child. I don't think that adults realize when they're teaching these stories that children are much smarter than they're being given credit for. Another of my many questions about the flood event was, how did all the animals make it to the ark, how were they fed, and what happened to them when they got off?

How they were fed was a question I was only beginning to acquire the math and biology skills to tackle, so I had to take it at face value that the ark was big enough to hold the animals and the food. It was later in my secular education that I gained enough knowledge to address this problem.

But, loving geography and animals, other questions were easy enough for a grade schooler to take on. And so I did.

Antarctica is an isolated continent, completely surrounded by water. How did the penguins make it all the way from there to the boat? That's a long swim.

Likewise for Australia and the kangaroos (I love kangaroos. I wanted one for a pet. My parents said no.) Did they somehow island hop halfway around the planet to catch their ride?

But somehow, magic happened, and all these critters made it to the ship, boarded, and had enough food on board to last an entire year. Then the flood waters receded.

The obvious question to me was, "How were the animals all going to get home again?" But I supposed if they got to the ark in one piece, they'd take the same route home. Then a bigger worry hit me: "What were they all going to eat?"

All life on the planet had been wiped out. There'd be no grassy plains for the herbivores. No oak or walnut trees for the squirrels.

But worst of all… what about the carnivores who lived together with the herbivores, who were an important part of their food chain?

Even at this point in my grade school education, I knew the difference between herbivores, carnivores, and omnivores. One thing from primary school biology classes was firmly fixed in my mind: carnivores could not digest plants. They had to have meat to survive.

The vision is as clear in my mind today as when it first came to me as a child:

Somewhere on a lonely African Savannah, the only two lions left in the world are staring hungrily at the only two remaining gazelles. If the gazelles get away, the lions starve to death and the species is wiped out. If the lions catch even *one* of the gazelles, that's the end of all gazelles—this was the last mating pair.

Maybe the gazelles are too weak to run at this point. The grasslands would have been wiped out after a year, so there'd be no food, and nowhere to hide. Maybe then gazelles just shrug their shoulders and give up. But we still have gazelles today. There would have been other animals for the lions to eat, but each time they fed, they would have wiped out a species.

Maybe a bunch of babies were born on the ark? The Bible doesn't say. But where would they have put them? The ship should have been full. And Young Earth Creationists were telling me that Noah only took babies aboard anyway, so they could all fit—they wouldn't have been able to reproduce.

No, wait, babies wouldn't work either. Living in Kentucky, I grew up around farms. I'd had basic biology classes and knew that mammals lived on mother's milk until weaned. So, there couldn't have been baby mammals on the boat. With no mothers, they would have starved.

In my mind's eye, the lions and gazelles just stand there on the savannah and stare at each other in utter confusion. How would this story end?

I brought up the basic biology and logistical problems with my church elders. As would always happen, I was told to deny what I'd learned in school, and to trust the Bible. Science didn't matter. Yet again, I was given a harsh warning that I was risking my soul just by having these thoughts. I was told to pray on the issue.

I felt enormous guilt over the thought, but it simply would not go away: What kind of plan *was* this, and who came up with it anyway?

"Punctured" Equilibrium

Evolution wasn't taught in rural Kentucky schools, other than a brief mention of Darwin and his finches. Teachers who attempted to teach this subject could, and did, receive threats from the public. Physical threats. Threats over their jobs. But yet, we children were sure we knew what it was. Well, we thought we did.

Everything we needed to know about the subject was taught in church.

In fundamentalist young earth teaching, evolution denied God as our creator. It claimed that we descended from monkeys. It may seem humorous to a scientifically literate reader today, but I literally heard the words spoken in church: "If we all came from monkeys, why are there still monkeys?" To believe in evolution was giving in to the Devil. Did any of us really believe that fish changed into dogs? And besides, the Earth was only 6,000 years old. How was that enough time for one kind of animal to turn into another kind of animal?

Amazingly, if you're brave enough to log into any online debate forum on the Internet today, you'll find these same arguments from fifty or sixty years ago still in use. As an information technology professional, it pains me to see evidence that all of our advances in computer technology have in some ways sent us backwards. An incredible amount of fundamentalist's evolutionary "education" comes only from Facebook, YouTube, and Sunday sermons. I wonder if the internet has advanced us as far as we'd like to think.

Anyway, back to the 1970s. One particular Sunday, our preacher was really excited. Science had proven evolution was wrong! See? He warned us!

Apparently, a "radical" scientist (who I now know was Stephen J. Gould) had proven Charles Darwin wrong! It didn't take long periods of time for new species (well, the preacher called them "kinds") to evolve after all. It could happen very rapidly in response to a catastrophic or significant change. The preacher called the new theory "Punctured Equilibrium." Yes, I too would laugh later in life when I took a real biology course and learned it was "Punctuated Equilibrium. But, the preacher was doing his best.

In my childhood, Charles Darwin was an evil man. They wouldn't tell us a lot of what he taught, just some "insignificant stuff about birds and turtles he found on some islands." But he was the very definition of evolution. And in fundamentalism, evolution contradicted our creator. To this very day, people who've never studied the subject still believe evolution revolves around Darwin, never realizing that the science has moved forward. Darwin was important, but DNA hadn't even been discovered during his time.

But, the preacher proudly proclaimed, the theory of evolution was now dead because Darwin was wrong, and… stay with me… this new "punctured" equilibrium thing proved beyond a shadow of a doubt that Noah's Flood was real! You see, because Noah could only fit a small number of animals on the ark, he was forced to just take the major "kinds" of animals. So, forget all the zebras, donkeys, and other creatures. Those all evolved from the horses Noah took with him. They were all the same "kind." Now that things were "punctured," it explained how the amazing different kinds of animals we see around us today all came from just a few pairs on board. They just quickly evolved, due to "punctured" equilibrium. "Cat kinds" quickly evolved into lions and tigers and mountain lions.

Wait. What? Our church believed in evolution now? I remember frowning and crossing my arms, with a perplexed look on my face. Was evolution real, or not? Just last Sunday, it was a tool of the devil. Now it was the foundation of today's sermon.

But it got more confusing.

One of the other prevailing arguments against Darwinian evolution floating around our church was the existence of the coelacanth. The coelacanth is an ancient fish. Darwin had to be wrong because coelacanths were considered to be living fossils. They hadn't changed over long periods of time, and evolution supposedly said everything changed. So either the Earth really wasn't as old as science claimed, or evolution wasn't for real. Or, both. (Unless it was really quick evolution that happened right after the ark hit dry ground.)

The upshot of all of this is that until this very strange and confusing sermon was preached, I didn't have an intense interest in evolution. Now, because I was one of those strange kids that went and looked everything up... well, I went and started looking things up. And it turns out that Darwin and Gould weren't necessarily going to be challenging each other to a celebrity boxing match to settle their differences after all (assuming they'd lived at the same time and knew how to box).

After a few hours with our encyclopedias at home, and some fresh science magazines from the local library, it became very clear to me that coelacanths had apparently never undergone any of the significant pressures that punctuated equilibrium predicted would cause them to evolve. In fact, it just seemed obvious that they hadn't undergone any pressures at all. There was no reason for them to change. My encyclopedias gave plenty of examples of species that *had* changed due to selective pressures.

Now, I'm not here to claim that as a youngster I could read and understand everything Gould was saying. But I can state with certainty that thanks to a traveling preacher who was misinterpreting something he probably saw on the news or read about in a Christian magazine, I had gained more knowledge than that preacher had. More importantly, he had given me the inspiration and motivation to start digging into evolution; ironically, the very thing he was trying to turn me away from.

I had friends from my childhood who tried drugs, alcohol, and cigarettes mainly because their parents tried so desperately hard to keep them away from these vices. Me? My congregation tried to keep me away from evolution because they considered it a vice. They ended up leading me right to evolutionary studies, and I found truth there.

Once again, I discovered I wasn't being given the full story at church. I went out into my garden of doubt and planted some more seeds.

Prayer

The car hit Michael with a sickening thud. Metal on soft tissue. I'll never forget the sight and sounds of the car striking him, of his body sliding down the street. He traveled perhaps thirty feet down the road in front of me, his head bouncing repeatedly off the pavement. A greasy trail of torn clothes, skin, and blood traced the path from the car's bumper to where his body now lay. A trickle of blood ran out of his head.

Michael was a younger cousin, about seven or eight years old at the time. I was six or seven years older. He and his sisters were visiting my family; the girls had walked down the country road on which we lived to our small neighborhood store to buy candy. In those days, it was very rare to see a car in our area unless it was a neighbor. As the girls walked back from the store, Michael saw them across the street. Excited, he ran to get his candy.

Except today, a woman happened to be driving down that road. She never had a chance to stop. Michael leaped out from behind a small pine tree, straight into the path of a car traveling around 40 miles per hour.

What happened next was confusing to me. Neighbors and family members ran to his body, held him, and began to pray. I was a devout Christian at the time, but, maybe as a sign that I naturally had a contrary way of thinking, I ran the opposite direction, into the house. 911 hadn't been rolled out nationwide yet, but I knew to dial 0 and get an operator (yes, I'm that old. Remember when we had operators?). She sounded a little skeptical about the story being told to her by an obviously young voice, but fortunately took my word for it and got medical help on the way.

I ran back to the street. My Mamaw (that's what we call grandmothers in Kentucky) rocked Michael slowly back and forth in her lap, blood running down her arms and hands, and pleaded with Jesus to save her grandson.

I shocked the adults by yelling at them. "Stop moving him! You could be making it worse!" No, I wasn't a doctor. Once, in second grade, a classmate had fallen six feet from a playground set, landing on her head and neck. Our teachers later turned the event into a teaching moment on first aid. You don't move accident victims. In my family, you never raised your voice to an adult, but, incredibly, they listened to me. How many times in my life would secular education step in and help? Everyone continued to pray, though.

In no way do I intend to ridicule or disparage them for praying. In times of trauma, I understand that such reactions are a way of dealing with tragedy. In the back of my mind, I think I prayed too. But to be perfectly honest, it was clear that action, not prayer, was necessary in this situation.

The ambulance finally arrived. Michael was in bad shape; he was stabilized and taken to a larger city hospital that could provide more advanced care. Our local country hospital wasn't up to the task. It took a long time, but he did eventually recover and, amazingly, except for a severe concussion, some broken bones, and a horrible road rash, he was none the worse for wear.

Nobody thanked me for calling the ambulance.

Nobody thanked the paramedics for stabilizing Michael and getting him 90 miles to the closest hospital that could save him.

Nobody thanked the doctors or nurses who provided his care.

Everyone thanked Jesus for saving him.

To be clear, neither I nor anyone involved wanted any gratitude for doing the right thing. What bothered me is that everyone was thanking someone who didn't seem to be involved, and everyone did the one thing that really wasn't providing any visible help at all—they were praying. I felt deeply blasphemous for the fleeting thought: Jesus could have prevented the accident in the first place.

I felt guilty that night when I said my prayers and asked God to forgive me for showing any disrespect toward praying. I did thank the Lord for saving my cousin. I knew in my heart, from church teaching, that I should have been giving credit to our creator for causing me to rush to the phone, and providing all the medical experts with the skills they needed to save Michael. To be honest, these thoughts rang hollow.

Throughout my life, I've seen countless cases where people prayed *after* a tragedy had already occurred. If prayer was going to help, why wasn't it being used to prevent the tragedy *before* it happened?

Conversely, if the prayer is "answered," people give credit to God. If the prayer isn't answered, it's simply explained away as "God's will." In fact, as we'll see in an upcoming chapter where I enter school as a pre-med student, the doctors are often blamed as failures. Prayer seemed like flipping a coin and calling it after you already know if the result was heads or tails.

I felt like a sinner for having the doubts that I describe here. I felt deeply guilty each and every time a rational thought crept into my mind and contradicted what my church was teaching me. I even felt that running for the phone, rather than staying to pray, might have been seen as a lack of belief in God's power.

Contradictions and conflicts like this would rack my mind, and my soul, throughout my childhood. I wanted so badly to believe. I was told that my doubts were the result of the devil tugging at me. But logic and science didn't feel like the devil. On more than one occasion, my fundamentalist congregation told me that science and logic were exactly that—Satan had a grip on me. As a young child, I was being confronted with a moral conflict that I don't think, in retrospect, any child would be prepared to handle: choose the reality I see in front of me and accept God's judgment, or ignore everything except what my church told me and gain eternal life.

Whore of Babylon

The Hippodrome Theater was a landmark in my small town of Corbin, Kentucky. The twenty-five-cent admission for Saturday matinees provided a great way for parents to offload the kids for a few hours of peace and quiet. I saw many memorable movies there.

Song of the South was a favorite cartoon for the kids, well before we were able to understand the racist overtones of the movie. *Bambi* was another crowd favorite, though to my dark sense of humor, the later classic short *Bambi Meets Godzilla* beat the original hands-down.

Diamonds are Forever turned me into a James Bond fan for life and cemented Sean Connery as the Best Bond Ever, though there is one who arguably could have contended for that title if he'd stayed in the role longer. I'm looking at you, Daniel Craig.

I remember staying curled up in my seat in horror during the debut of *Jaws* (I was only 10 years old). But I was still hooked. "You're gonna need a bigger boat!"

However, the most horrifying movie I ever saw at the Hippodrome was recommended by my church. The movie was a visualization of the book of Revelation. I've tried in vain to find the title of the movie. Christian-end times movies were very popular in the 1970s and 1980s, so there are many possibilities, however, I can't quite pin down this one. Most likely, it was a low-budget film produced by evangelicals and didn't receive a lot of attention. But it was touring theaters in the Bible Belt; our church heard about it, and they encouraged everyone to attend. In fact, nearly all churches in town were sending their children. The movie was so well viewed that it was a topic of discussion in our secular school classrooms; even on our playgrounds.

And so it was that late one Fall evening, I found myself in attendance with two of my cousins, all of us teenagers, to learn what was going to happen at the end of the world.

I was unbaptized (my Catholic baptism didn't count, remember) at this time, so I walked in well aware that the sinners who were going to be punished included me.

The movie was a blur of vignettes of people suffering. The title character of this chapter, the Whore of Babylon, will stay in my memory forever. She was a beautiful woman in a purple dress. She held a golden goblet in her hand. When she drank from it, blood poured out of her mouth. The narrator explained that this was the blood of Christian saints and martyrs.

I didn't stop to ask why God hadn't protected these Christians from having their blood poured into a cup for a whore to drink. At the time, I wasn't exactly sure what a whore really was. I led a pretty sheltered life. I just knew it meant a woman who did "bad things." The particular woman terrified me. I didn't want my blood to be in somebody's cup. I felt sick to my stomach. I put down my popcorn and soft drink and left them. This was no time to eat.

Next came a screen full of shadowy human figures, all writhing and screaming as they burned in flames that took up most of the screen. This was the punishment that they *deserved*, we were told. I'd been indoctrinated well enough by now to understand that people brought this upon themselves because they hadn't accepted God and followed the teachings of Jesus.

I curled up in my seat in terror. Children throughout the theater were crying.

The narrator mentioned the necessity of being "washed in the blood of the lamb, Jesus" to be saved from this fire. Wait. What? The whore was just spewing blood, and now we're going to have to be covered in blood to escape the fire?

My cousin leaned over and whispered, "He means getting baptized." Oh, right.

The people behind us told us to shush and stop talking during the movie. It is worth repeating that the majority of those in attendance this night were in fact young children. I wasn't the only one talking. I could hear gasps and cries throughout the audience.

I was deeply disturbed. I didn't want to burn. I didn't want those I loved to burn. The narrator continued to intone that we all had an easy way out of all the horrors shown on the screen. The choice was all ours. Yes, there would be trials and tribulations along the way. But if we accepted Jesus and stayed true to our faith, the terrible, eternal punishment could be avoided.

One other part of the movie that would stick with me forever was the way God was going to announce the onset of the end of our world: an angel was going to appear and blow a loud horn. Everyone would hear this and know it was the beginning of the end.

I just happened to live near an interstate for most of my life. Trucks blew their horns during all hours of the day and night. From that evening forward, I'd lie in bed, visions of burning people and blood-spewing whores tormenting my thoughts. Every time a truck horn would blow on the interstate, if I was awake, I'd pull the covers over my head and plead, "Oh God no, is this it?"

It was worse when I was sleeping. The horns never failed to startle me awake.

I'd sit up in bed, crying and screaming. The brimstone lake terrified me. I didn't want a whore to drink my blood. The pressure of countless Sundays of learning that I was flawed, that I was giving in to evil by having secular thoughts… it weighed me down as if I was holding a heavy stone in a deep lake. Even my waking thoughts were troubled. I'd sometimes be found trembling in fear after sinking deep into thought about my future if I wasn't "saved."

It may be hard to imagine that this all started with something as benign as me openly doubting the origin of rainbows in Sunday school. But it did. And it got worse from there. Fundamentalism taught me that every word of scripture was true; nothing could be doubted or denied. The nightmares wouldn't stop. People at church told me there was only one way to avoid it.

I had a decision to make.

Immersion

The pressure had finally become too much to bear. Visions of burning in Hell. The screams. Skin peeling off and being regenerated. The recurring nightmares. Truck horns on the interstate in the middle of the night shattering my sleep because I thought the Seven Angels had returned to signal the end of the world.

One morning at the end of Sunday school, as the rest of the children filtered down the stairs for regular church service, I stayed behind. I meekly walked up to my teacher and announced that I wanted to be baptized "the right way."

Recall that I had been repeatedly told that my Catholic baptism carried no weight in God's eyes, because I hadn't been totally immersed. It would be years before I began language studies and learned that my church was mistranslating Greek, and my first baptism was perfectly in line with the definition. My journey to reason would involve slowly stripping away layers of misinformation from my childhood, but this layer would have to wait a few years.

My teacher was ecstatic! She smiled and hugged me with tears in her eyes. She led me down the stairs and said, "You're going to sit with me for the service." I caught her winking at my father and sisters several pews back. Mom, being Catholic, attended a different church, so she wasn't in the audience.

I don't remember much of the sermon that day. I was too focused on what was about to happen. I'm terrified of water. Like every evangelical church, ours had a large deep baptismal in the front. I stared at the water the whole time. Oh God, I had to go in there.

If you're not familiar with evangelical services, they all end with the crowd rising to their feet and receiving an invitation to come forward and be saved. This is known as an altar call. A particularly long hymn is sung, and to make sure everyone has a chance to give in to the urge to step forward, they sing every single verse. All five of them when the song allows it. The preachers stand there with hopeful looks on their faces, eyes wandering over the crowd. "Step forward, step forward!" seemed to be their telepathic message.

So it was during this final hymn that I felt my teacher take me by the shoulders and lead me to the front of the church. I was extremely nervous. I hadn't talked to my mom or dad about this. I was fully aware I was openly renouncing my mother's religion. A small gasp went up from the crowd. No doubt many of those sounds were of delight, yet the cynical adult version of me can't help but wonder if there were some "Oh damn, now dinner in the crock pot is going to be overcooked because we have to sit through a baptism." Or, for some, "Well, I'm going to miss the NFL kickoff now." (VCRs weren't affordable in those days, so Sunday after church was always a somewhat controlled rush to say goodbye and get home to eat and be settled in for the 1:00pm kickoffs.)

Sorry, I digress. The preacher was thrilled. He had the look of a fisherman who'd been at it for weeks and finally reeled in a large fish.

Many words were said, none of which I particularly remember. The crowd was reseated and I was led back upstairs by the preacher to the Sunday school classrooms to change. He handed me a robe and asked me to strip to my underwear.

Oh no, not again. Alone with a man who was asking me to undress, just like the babysitter did years earlier. I'm not sure that the trauma of sexual abuse ever really goes away. As my life progressed, I'd come to feel the same way about the suffering incurred via fundamentalist indoctrination.

I mustered the courage to ask the preacher to wait outside. I didn't explain why. I think he mistook it for modesty and kindly agreed. The thought occurred to me, maybe this baptism would cleanse me of the sins I thought I'd committed with the pedophile babysitter.

I got the robe on and announced I was ready. The set of stairs that led to the baptismal were accessed through a door in the classrooms, so the preacher opened the door and down the stairs we went.

The water was chest high and *cold*. I wanted to turn back but the man was behind me, guiding me down the steps. Panic set in.

The preacher had me pinch my nose tight and after a quick "in the name of the Father, the Son, and the Holy Ghost," I was plunged backward into the water. I freaked out. I tried to scream, which is never a good idea underwater. I tried to fight back. I landed several solid punches on my attacker's chest, but my arms weren't long enough to reach his head, where I could do real damage. It seemed like I spent an eternity underwater, though of course, the reality was that it was a quick plunge.

I was finally scooped out of the water, choking, by a smiling man whose face was well within reach of one of the roundhouses I was trying to land while he held me under. But I stopped. There was a wall of sound. I looked out at the audience. They were all on their feet, clapping, stomping, and cheering.

I'd done it. I was a real Christian now.

I went upstairs and changed back into my church clothes. My wet underwear was causing a lot of problems that only other guys can really understand, but since I had made a snap decision to get baptized, I never really thought far enough ahead to bring a spare pair. They never warn you about this in church. To add to the discomfort (and further upset those who were hungry and awaiting NFL kickoff), the whole church was asked to remain behind so I could get changed and meet them at the door to be congratulated as they left. Everyone seemed very happy for me, though I could swear I heard stomachs growling.

We lived only three houses from our church, on a quiet country road. I remember walking home alone that day, hair and underwear soaking wet. Looking back, I think Dad went ahead to let Mom know what happened. Remember, for them to be allowed to marry, they both had to promise the Catholic Church that their children were to be raised Catholic. In my baptism, I'd just caused them to break a guarantee made to Mom's family, her church, and to God. While I was overjoyed to finally be a "real" Christian, I also felt like a traitor to my mother and father, and the oaths they had to take in order to be married.

To my relief, when I got home, Mom was fine. She'd seen all the struggles I was having with religion and I suppose she felt that this might bring me comfort. In all honesty, as time went on, I came to learn that my parents didn't care so much what religion we followed, just as long as we were good kids who followed God.

I think the hope among my family was that this baptism might put to rest the problems that religion was causing in my life.

They couldn't have been more wrong.

Parenthood

One question that would come to my mind if I were reading a book such as this is, "With church causing all this turmoil in your life, what did your parents do about it? Didn't they see the problems? What did they do to help?"

This is a fair question. My parents did see my pain. And they intervened early and often. Before I explain in depth, though, I think it's important to set the context in terms of how the world (in my experience) views religion. I must also point out here that my parents would turn out to be among the strongest positive Christian influences in my life. Not by their faith, but by their deeds.

Good parents (and I'd argue that mine were the best) care about their children. They strive to provide them the things in life they need to survive. Not just food, shelter, and clothing, but a good education, and a life that teaches them moral and social values that will make them better people.

But, we happen to live in a world where billions of people think that religion is the absolute definition of good. It's no coincidence that most children in Muslim countries grow up to be Muslims; most in predominantly Christian countries grow up to be Christians, and those in Jewish communities grow up to be Jewish. I'm not ignoring other important religions—I'm sure you see my point.

We're all born atheists, but our parents, if they happen to be religious, see fit to make sure we're brought up in the religion that *they* were taught was the source of good morals. In their eyes, they are doing their jobs.

And so, for me, the religious choice was Christianity. Never mind the conflicts between the Catholic and Protestant denominations—I was being brought up to believe in Jesus. Yes, there was consternation in the Catholic half of my family that I wasn't going to be a Catholic, and joy in the Protestant half that I was no longer Catholic, but at least I wasn't a member of those *other* religions. In fact, one of my three sisters did go on to become fully Catholic, much to the joy of half the family. But the other side? You just can't make everyone happy.

But me... I was a Protestant. A fundamentalist evangelical Protestant. A member of a church that brooked no compromise. As I've stated elsewhere, I was frequently taken aside by members of my church and told, "You really need to do something about your mother. Catholics aren't true Christians. It's up to you to save her."

In fact, after my baptism by total immersion, the nightmares involving my mother worsened. How could I possibly go to Heaven without her? The recurring dream, constantly reinforced by my Sunday school teacher telling me Mom was going to Hell: I would literally be on a cloud in the sky, enjoying the ability my new angel wings gave me. I could fly! But then I would hear horrible screams. Looking down, I could see people burning, just like in the Whore of Babylon movie. One of those on fire was my mother. We would look into each other's eyes. Flames enveloped her body, just as my church described. In agony, she would plead for me to go talk to God and apologize for her rejecting the true religion. However, God always said "No. She had her chance." I'd return to the edge of my cloud and, sobbing uncontrollably, explain to her that it was too late. My apologies for failing her, for not converting her, were still on my lips as I'd wake up, moaning, reaching out in vain to pull my mother out of that lake of fire.

My church told me this was going to happen, and I believed them.

I admit I did not fit the mold of a normal child. I went "all in" on everything. Given a school assignment to produce a two-page paper on the planets, I wrote an entire spiral-bound notebook. Told that not everyone on Earth spoke English, I dove into foreign language books. Claimed by my church that I was in the one true religion and all others were false, I went headfirst into Christianity, reading the Bible cover to cover at least three times. When I didn't want to eat my broccoli and was informed there were starving children in Africa that would give anything for that meal, I found a box and asked how much of my allowance it would take to send my dinner overseas.

Yes, that last incident really did happen, but is perhaps a digression. What I'm trying to convey, with apologies to the reader for even the slightest hint of egotism, is that I was somewhat of a forward-thinking kid, with the ability to learn rapidly, and to reason in ways that sometimes confounded adults. But I don't think I'm special in this way. I fear that many young children, currently undergoing indoctrination, may be struggling with the same feelings.

So, my parents did comfort me when they saw the conflicts. They patiently explained that my mother and other Catholic family members really were Christians, and I didn't need to be upset by the things I heard at church. I don't know if they ever specifically spoke to any of the members who were pressuring me. I suspect they did, but once you've gone all-in on fundamental Christianity, it's very hard to see other viewpoints. I was now torn between two worlds: the reassuring words of my parents and the teachings of my church.

One of the major problems is that I did not necessarily want to be talked down from the high, theologically superior perch my fundamentalist teachers had placed me on. I wanted the salvation they promised me, and I was so fully indoctrinated that I was absolutely convinced there was only one way for me, and everyone else, to get there. I'm sure most if not all parents have been faced with the problem of trying to convince their children that there's not necessarily only one viewpoint—the child's/teen's—in the world. Good luck with that.

As mentioned, Mom and Dad were undoubtedly at the top of my list of good Christian role models. They didn't preach their religion. They lived it. They were the first to volunteer when a family member, or even a stranger, needed help. They didn't just warn their children not to engage in immoral behaviors, they avoided those behaviors themselves. I could see their religion in action. Thinking back to how Dad handled the episode where I committed theft as a young boy, he did it with Christian forgiveness while still managing to teach me right from wrong.

But there are two forces at work here: what God supposedly says, and what your parents are saying. Unfortunately, for all the good my parents did, for all the patient and loving talks they had with me whenever I came to them, I was fluent in the Bible, and my fundamentalist congregation hammered home the following scripture, early and often:

Matthew 10:35-38

"35 For I have come to turn a man against his father, a daughter against her mother, a daughter-in-law against her mother-in-law.

36 A man's enemies will be the members of his own household.

37 Anyone who loves their father or mother more than me is not worthy of me; anyone who loves their son or daughter more than me is not worthy of me.

38 Whoever does not take up their cross and follow me is not worthy of me."

Yes, my parents did intervene. They did notice what was going on. They lovingly counseled me whenever I was troubled. But they didn't understand the true power the church had over me. Like so many parents today, though, they kept me in religion with the misbelief that it was for the greater good. It just so happened that my congregation taught me that their word—God's word— superseded the teachings of my mom and dad. Had I been able to make the connection with what I'd learned earlier from interviews of surviving family members from Jonestown, I would have recognized a common thread. Believe only in your religious teachings. Outside influences are to be disregarded. This is dangerous thinking. I did not recognize it at the time. And so, my father and mother fought a losing battle.

Who was I going to trust? My parents or God? My church told me it had to be the latter.

The Gorilla

Milwaukee is my true hometown. Even after moving to small-town Kentucky as a child, our family would make yearly trips home to visit relatives. I loved Milwaukee. Living in a small southern town with a population of 10,000, while it had its benefits, didn't really offer a lot of entertainment or educational opportunities.

One of my favorite Wisconsin pastimes, other than Summerfest and the State Fair, was visits to the city's zoo. To a Kentucky kid whose daily encounters with animals were limited to horses, dogs, cows, and chickens, the zoo was a wonder world.

One particular event at the city zoo would plant another seed in the garden of doubt that was continually being fertilized by my church's insistence on a 6,000-year-old Earth, with all of the animals' existence explained by a trip on a large boat.

It happened in the primate exhibit. At the time, zoos were not particularly enlightened, and there's a justifiable debate over whether or not even modern institutions are ethically treating animals. But this was in the 1970s-1980s, when such arguments weren't much in the public mind.

There I was in the exhibit, when I noticed a gorilla in a glass cage that I'd estimate to be no more than fifteen by fifteen feet. The floor was concrete, painted what I would come to refer to as "puke green." The only contents of this cell were a rubber tire hanging from a rope tied to the ceiling, and bowls for food and water. Excrement covered the floor. A small hole covered by a grill was in the center of the cell, but of course the gorilla was not trained to use it. It was obviously there for keepers to eventually wash the mess down the drain when they got around to it.

I walked up to the glass and pressed my nose against it. The gorilla came over and joined me. He pressed his nose to the glass too. I put my hand on the glass. He matched my movement. We stared at each other.

That's when I saw it. My own face, reflected in the glass, superimposed over the face of the gorilla. I gasped.

We looked the same.

Yes, yes, he had a lot more hair, and his head and body didn't match mine exactly, *but we were essentially the same.* Maybe not brothers, but perhaps that one cousin that nobody in the family talked much about.

Please understand, in Kentucky grade schools and high schools, evolution was a taboo topic. In this time period, teachers who tried to do justice to the subject were essentially labeled "ugly atheists" and run out of their jobs (and towns). I wouldn't learn this until later in life… I'm just pointing out I had no concept of evolution, other than what had been taught in church.

And those teachings weren't pretty. And, as it turned out, they were inaccurate.

"Evolution teaches that humans came from monkeys," my church told me. "There are plenty of monkeys around today. Look at them. Do you think they gave birth to a human?"

I can feel the scientifically educated readers out there shaking their heads… "Did they ever really teach that?"

Why yes, as a matter of fact, they did. And they still do. Hop into any modern evolution debate forum and you'll still see this question, verbatim. I promised this book would give insight into why some Christians today behave the way they do. Opposition to evolution? Ridiculous human-from-monkey arguments? Look no further than your local evangelical church. Of course, many denominations have embraced evolution and have no major conflict with the science behind it. But visit the Bible Belt. Ask a local school board why this subject isn't taught in schools. And be prepared to run.

So, the gorilla and I continued to stare at each other. To a fairly intelligent teenager, it was clear that there were some serious gaps in my education. Despite all my indoctrination, I realized I was looking at something significant.

I understand the dangers of assigning human emotions to other animals, but I couldn't escape the sensation that this gorilla was very sad. Trapped in a small, excrement-covered cell with nothing but an old tire to play with, his life was going nowhere. There I was, a creature that looked hauntingly like him, yet I was free to traipse through the exhibit and enjoy an ice cream afterwards. It was like visiting a relative in prison.

I don't know how long we stared at each other. I resolved then and there that there were things I wasn't being told. Things I needed to learn. There were too many similarities here. I felt very strongly I was not looking at a coincidence.

My reverie was interrupted by a family of humans who came up and started banging on the glass, making monkey sounds, hopping around and scratching their armpits. The gorilla quickly retreated to a corner, visibly distressed. I yelled at the humans. My religion taught me that cursing was wrong, but I let slip some words my parents didn't know that I knew.

I'd pray for forgiveness later. At that moment, the family of glass-bangers deserved every critical word that came out of my mouth.

Regardless, I had homework to do. This was, again, in the days before the World Wide Web, but as I've mentioned, my parents cared enough about their children's education to maintain two complete multi-volume encyclopedias at home. One was a children's encyclopedia that didn't tackle evolution at all (though I did learn to make gunpowder from it). However, the adult volumes tackled the subject head-on, and it was an eye-opener.

Evolution is a complex topic and I won't claim I came fully to grips with it after several hours of reading. But I did learn why the gorilla looked like me. He was a Great Ape. And so was I. My instincts from the zoo *were* correct. We *were* related. Neither of us came from a monkey. We just had a common ancestor. It was so easy to understand.

Why was this a problem? Why was my church covering this up? Why were they lying about it? Why were my biology classes in school not talking about this?

I knew not to bring the subject up in church. By my teenage years, church was no longer providing answers that I couldn't destroy based on the secular education I was receiving. Because of the pressure I received from my congregation, I didn't *want* to tear the church's teachings apart. I still believed in God. I still believed in Jesus. I just needed answers that didn't contradict what reality was showing me. I was a fully-baptized, born-again Christian. I felt guilty for disobeying God by believing that the gorilla and I could be related.

Time and again, my secular education would clash with the theistic, leaving me struggling to do the impossible: push scientific thinking to the back of my mind.

I do recall broaching the subject in biology class. As I've said before, I feel sorry for teachers in the Bible Belt. Today, as was the case when I attended school in the 1970s and 1980s, educators are constrained by what they can say—either by the inevitable backlash from parents or, now, sadly, by actual laws that prohibit them from speaking the truth.

And so, my biology teacher couldn't (or wouldn't) help me either. He had an embarrassed look on his face, and the class went quiet over the audacity of my suggestion that humans were related to other creatures. I'd just asked my teacher a question that could literally cost him his job.

And, at recess, my fundamentalist classmates danced around me and made monkey sounds.

Easter

I look back with shock and embarrassment over my early church education on Judaism. Contrary to current-day views toward Israel and Jews, evangelical churches of my time were extremely antagonistic toward this religion. "The Jews killed Jesus" was a theme of every Easter sermon I ever attended. And I mean all of them.

Antisemitism wasn't new; it was firmly established in history long before I was born. A common slander against Jews was *blood libel* that they used the blood of Christian children for ritual purposes. Those who have studied the Holocaust know this was a charge leveled against Jews in Germany. Given that the last pre-war census in that country had it between 95-97% Christian, I can look back at my fundamentalist church and see that its antisemitic roots had been carried forward from history.

Being Bible-literate at a young age made me recoil at the annual Easter preaching against the Jews. It was the Romans who crucified Christ. Crucifixion was a Roman torture, not a Jewish one. Once again, this was my secular education speaking to me. My Bible Belt schools seemed to carefully avoid American slavery, but Roman slave revolts, and the punishment for the slaves, were covered in classrooms.

On Easter though, our church would always trot out Matthew 27:24-25, where Pilate literally washes his hands of the condemnation of Jesus, and the crowd allegedly declares, "His blood is on us and our children." I say allegedly because, after learning later in life when the Gospels were actually written and the religious motivations of those who wrote them, it seems apparent to me that Jews, who had rejected the teachings of Jesus, were made scapegoats early and often.

At the time, I thought it telling that not a single follower of Jesus came forward to defend him. I still find it odd.

But, in a day long yet in my future, when I saw a crowd of right-wing zealots marching through the streets of Virginia chanting, "The Jews will not replace us," I sensed where their education came from. It was a flashback to the teachings of my youth. Apparently, with antisemitism on the rise, people are still forgetting that it was the Romans who carried out this gruesome act.

Amazingly, as I grew older and the age of televangelism took off in the 1980s, I noticed a surprising change of course in evangelical Christian churches. The Jews were no longer the enemy. Pat Robertson and his colleagues were actually telling us we should support Israel and the Jews. Apparently, the rise of Israel was the fulfillment of a prophecy foretold in Ezekiel. Churches in my time began preaching of the Jews as God's chosen people, the necessity for all of Jerusalem to be reunited, and the rebuilding of the ancient temple (where now, problematically, a mosque now stood). Muslims would become the new adversary.

Even though antisemitism never went away as the years passed by, conservative Christian support for Israel grew, and started making its way into government policy. Because, you know, a large part of the government leadership positions were held by conservative Christians. It actually became illegal at one point for American businesses to protest, via boycott, Israeli actions in the disputed West Bank.

So, I never understood, and probably never will understand, the Christian position(s) toward Judaism.

I do know that I eventually came to admire the religion so much that as an adult, I would begin conversion classes.

Tabloids

I developed a sense of inquiry and skepticism early in life, and one of my motivations was an unusual source: my deeply religious Catholic grandmother. Grandma was a huge fan of tabloids such as the National Enquirer and the Globe. She had weekly subscriptions to three or four of these "papers," and, I must admit, she believed a great deal of what she read in them. Ancient aliens, celebrity gossip, horoscopes… much of this was real news to her.

I ask the reader not to judge my grandma too harshly for this. She was the daughter of a recently-arrived Polish immigrant, and didn't have access to the educational opportunities I had. As was the custom in the early twentieth century, she married young, had children when she was young, and was left to take care of those children by herself when their father, an alcoholic, left home. She was one of the kindest, most loving persons to have ever walked the face of this planet, and I loved her dearly. The fact that she might be a little… "susceptible to suggestion" …it just made her all the more adorable.

One of grandma's favorite sayings was, "See a penny, pick it up, and all day long you'll have good luck." For better or worse, she actually believed in this. During one of our yearly family visits to Milwaukee, my prankster dad started leaving pennies lying around the house. We eventually found Grandma limping up the stairs, a loud metallic clanking sound coming from her right shoe. Apparently, when you picked up a good luck penny, it had to go into your shoe to make it work. Grandma was a good sport though.

She laughed, told my father he wasn't getting his money back, and put it in her bingo jar.

Grandma was one of the positive Christian influences in my life. She was full of forgiveness, always saw the good in people, and never held a grudge. Yes, she was perhaps a bit lovingly gullible. It's not surprising she believed the things she read in her tabloids (or the Bible).

I'll now ask the reader not to judge *me* too harshly, because I also loved the tabloids. I dove into them as soon as I could read. Grandma always saved every issue for my visits. I didn't enjoy them for the same reason, though. Being one of those annoying kids who questioned everything and needed convincing about every single word I was told (like, "Eat your broccoli, it's good for you."), tabloids were fertile ground for me.

I had a very skeptical mind but was never allowed by my church to apply it openly to scripture.

But I was allowed to question tabloids, which had much in common with the supernatural teachings I'd encounter in church. In a sense, the National Enquirer et al. became safe proxies for my scientific criticisms. Questioning a supernatural event in holy text was expressly taboo. Examining the same situation in the World News was fair game.

For example, it always struck me as odd that the Bible talked about dead people rising and being encountered on the streets. But it was deeply ingrained in me that I should accept this on faith. And so, I did. Concerning holy books, I was, I'd much later realize, programmed to accept what I was told at face value. I completely and wholeheartedly accepted that Jesus rose from the dead, and that he was able to resurrect others.

But Jesus, and my church, weren't around when I was reading

the National Enquirer. At least not to my naked eye. And so, question Elvis' resurrection, I did.

Reading and debunking tabloids was one of my early hobbies. It was actually enjoyable, as strange as that might sound. Being allowed to openly question things without fear of retribution was a breath of fresh air.

Noah's Ark seemed to be found in a new place at least once a month. I always struggled accepting that the Earth had once been covered in a flood, and, as mentioned earlier, was deeply troubled by innocent children being drowned. I was taught to tuck those doubts away, and for a while, I did. But all these ark discoveries? This I could question!

So, OK, maybe there was one Noah's Ark… but a new one in a different location every time the printing press ran off a new copy of the Enquirer? Why was the wood in the photo not rotten by now? Why were modern nails sticking out of the wood? Why was I being shown just one small scrap of lumber when it was supposed to be the whole ship? How did they know that material came from a boat? What if it was just a piece of wagon?

Astrology was one of my favorite targets. I loved *astronomy*, a real science, since the first time I remember seeing the night sky. Astrology, though, was a different matter. Even my church called astrology evil, so I was free to nitpick. How in the world could, well, other *worlds*, affect me here on Earth?

A sample horoscope from my youth read, "Your love life will be colorful and vibrant today."

Say what? I hadn't gone through puberty yet when I first saw something like this. I had no love life. But even if I did, what did "vibrant" mean? I had to look it up in a dictionary. Yes, of course, I did stumble across "vibrator" during this visit with Merriam-

Webster, but I was too young to get the dirty joke.

It finally occurred to me to open up several different tabloids at the same time and compare all the horoscopes to each other. They all couldn't have been more different! This was clearly utter nonsense, and the fact couldn't have been more obvious to even a young child. I was flabbergasted that actual adults believed this stuff.

In my eyes, the vague wording of horoscopes also applied to words of so-called prophets, like Nostradamus. Predicting wars is not a hard thing to do. I grew up with Vietnam on the television every day. Before my eleventh birthday, Israel had been invaded twice. My uncles fought in the Korean War. My grandpa and older uncles fought in the Second World War. History class at school was a litany of war after war after war.

But when Nostradamus predicted a war, it made headlines. It was painfully obvious to me that every war he "predicted" just so happened to be whichever war was currently going on.

I started to annoy friends and family who read their horoscopes every day by making my own predictions:

"There will be an earthquake this year."

"There will be a flood this Spring."

I was *always* right. Every prediction I made came true. As a family, we would watch the CBS News every night, whether we were in Wisconsin, Kentucky, or elsewhere. I saw my predictions come to life on television. Forgive me, but I rubbed it in with those who I knew believed in astrology and prophecies (well, as long as they weren't *church* prophecies).

"Hey cousin Sue, remember that earthquake I predicted? Well, it just happened. Saw it on the news last night. What's your horoscope predicting for today? Bet I can come up with one just as good!"

"Shut up, Mark. I'll believe what I want to believe. Just shut up."

Obviously, in hindsight, debunking tabloids is picking low-hanging fruit. Remember, though, I was a child at this point. And I was forbidden from questioning the supernatural events of scripture, no matter how parallel the themes often seemed to be. If the contradictions were so obvious to a kid with a grade or middle school education, why weren't they so to grownups?

Why oh why could nobody get a clear, steady, in-focus picture when photographing Bigfoot or flying saucers?

I marveled at how many times Jesus or the Virgin Mary were found in burnt toast or other food products. I looked at the photos and, with a lot of imagination, maybe... *maybe*... I could see something. But I had to work hard at it. Why were these beings appearing to us in bread instead of in person? I started intentionally overcooking or burning my toast to see if I could make Jesus or Mary appear to me. Given that we were not a very wealthy family, my mother did not appreciate this... food was something that wasn't to be wasted. And yet, sometimes I succeeded. If I burned enough toast, I could get an eerie human figure. I was told the Enquirer would pay good money for a story. I saw income potential. My dollar-a-week allowance needed supplementing. Unfortunately, at this time in history, without the internet, I didn't have a way to market my creations. **

Aliens were a constant feature of the tabloids. 1970s America was UFO-crazy. On this topic, I *wanted* to believe. I had finished Isaac Asimov's *Foundation Trilogy* somewhere between ages twelve and thirteen, and it opened my imagination like no other book before. Races of beings scattered across countless planets was a vision that made sense to me. My church forbade thought and discussion on the subject—God had created us and us alone, a Sunday school teacher explained. If he had made other planets and other beings like us, it meant they were all going to burn in Hell, because Jesus only came to Earth, so he wouldn't have been able to "save" anyone on another planet.

(I knew not to discuss Asimov at church. He was an atheist. All atheists were evil and everything they said was meant to mislead us. Isaac Asimov's books were not allowed in our school library. *How could anyone tell me what books I wasn't allowed to read?!* It took a long time to get my hands on the Foundation Trilogy. But I still have a copy, and re-read it every few years. What an amazing work.)

But speculate on aliens, I did. Extrasolar planets were unknown at this time. It would be decades before we began discovering them by the thousands. Today, we know that many, if not most, stars have planets. We didn't know that then. While the thought of other planets (and perhaps life) was ridiculed in evangelical churches, the tabloids—and the imagination of the American public—ran wild with the thought.

I had questions, though. How were the aliens getting here? I was fortunate enough to have a sixth-grade teacher who did a good job of breaking down the very basics of Einstein to his science class. I understood the speed of light was a maximum speed limit. I knew from my burgeoning (self-taught) astronomy education how far away the stars were. Even at this early age, it was easy to grasp how impossibly fast aliens from even the nearest star systems would have had to travel to reach us. In 1969, my very excited father pulled me to the television to watch the first landing of men on the Moon. He correctly predicted this would be a major story in the history of humanity, and wanted to make sure I saw it.

"Remember this, it will be important," my father told me.

It was just within my mathematical capabilities at age twelve to compare the speed of an Apollo 11 rocket to the speed of light. I did my best, and went to a teacher who, as so many of our great, dedicated teachers still do, was delighted to help. The answer was astonishing. Expressed as a percentage, humanity's approach to the speed of light had so many zeroes between the decimal point and the first non-zero digit that it was essentially zero when rounded to any reasonable limit.

So maybe humans couldn't pull this off, but, as the Globe would declare, these beings visiting Earth were obviously more advanced than humans. *They were finding ways.*

OK, fine. The aliens were so advanced that they'd figured out problems that made my childhood head spin. But if they were so advanced, why in the world couldn't they keep their spaceships hidden once they arrived here? That would seem a trivial task. And, more striking, why did their advanced ships keep crashing all over the place? Were designers of interstellar spaceships, machines so complex that they were beyond the abilities of humans, so incompetent that they couldn't stay invisible when they arrived at Earth? I was a big Star Trek fan in the 1960s. Apparently, I was biased by the thought that the Prime Directive made sense and aliens would be abiding by it. Yet there they were, popping up next to airplanes, hovering over buildings, kidnapping cows… it didn't make sense.

To wrap this all up: tabloid newspapers might not seem to be the best educational material, but to a fertile childhood mind that loved to ask questions, they were a great training tool for rational and scientific skepticism. They were presenting incredible claims, just like my church was. I was met with a harsh response whenever I openly questioned a church teaching. These "newspapers" became an outlet for practicing critical examination, and helped avoid the frustration I experienced in not being able to question theology without threats of eternal torture.

Much later in life, when I finally felt the freedom to apply the same type of skepticism to church teachings (e.g., talking snakes) without fear of reprisal, my life would change forever. In these distant future moments, I would always find myself going back to the thought in the back of my mind whenever I held a National Enquirer in my hands:

"If a kid can pick these arguments apart, why can't an adult?"

** Many years later, in 2004, after the invention of the Internet and World Wide Web, a piece of toast featuring an "image" of the Virgin Mary sold for $28,000 on eBay. I had truly missed my calling.

Busha

My family is extremely proud of our Polish heritage. Over the course of generations, we have achieved the proverbial American Dream. It started with my great-grandmother, always known to us by the Polish word for grandmother— "Busha"—leaving Poland in the early 20th century, and evolving into a host of great-grandchildren who took full advantage of what our country had to offer. I was the first on either side of the family to graduate from college, but two sisters would surpass me, earning Masters and then Doctorates. Busha could have only dreamed of such an impossible future when her ship came to Ellis Island about a decade before the Nazis would turn our home country into a horror show.

If not for my father, that "damned Kentuckian," my sisters and I would have had a full unbroken bloodline of pure Polish lineage going back at least one hundred years. Dad messed all of that up with his English/Scottish blood. But this still made all of us the descendants of immigrants, some very recent. Of course, we all felt totally American, and we *were*, but our roots still meant a lot to us. Busha spoke very little English. We'd huddle at her feet as children, listening to her stories of Warsaw and the Old World as our mother and grandmother translated for her. We used the little Polish we knew and Busha would beam with pride and pleasure. She pinched our cheeks. For some reason, Polish women always pinch cheeks. My grand-nieces and grand-nephews still cover their cheeks with their hands when they see my mom coming. Trust me, it's just part of us.

Being Polish was no big thing in Milwaukee or nearby Chicago. There were plenty of us there. We were a community. But when we moved to Kentucky, things changed.

As I've mentioned, the first hint I got that I was different came from church. Poland was, and is, almost entirely Roman Catholic. That's the faith I into which I was originally baptized. My parents had to promise to keep their children in this faith before marriage was permitted. Keeping that vow when we moved to small-town Kentucky was hard. There was one Catholic church, and countless evangelical protestant churches. Catholics were viewed with suspicion.

When we began visiting the Young Earth Creationist church my father had attended when he was a child, it was more a matter of convenience than anything else. This church was within walking distance of our new house in Kentucky. The Catholic church was on the other side of town. My parents didn't much care which church we attended, to be honest. They cared more that we just grew up as Christians… didn't matter what kind. Neither Mom nor Dad completely subscribed to the rigid teachings of their parent churches. I'm not sure they realized the affect the doctrines of those churches would have on young minds.

As I've said ad nauseum, my new church quickly taught me that I was not a true Christian at all. My most serious problem was that I hadn't been baptized by being fully immersed. Water sprinkled over the head just wasn't a proper baptism. And then there was the problem of worshipping all those idols (statues) and the pope. Praying to saints was sinful. We were supposed to be talking directly to Jesus or God.

But it wasn't just that I wasn't a true Christian. I wasn't a real American. I would proudly tell people in my church that I was Polish. I could even speak some Polish for them!

"Oh no, Mark. You are an American. See that flag up on the altar?" They'd point to the American flag standing next to the

Christian flag in the front of the church.

"A lot of Americans died so that we could worship Jesus. Those red stripes on that flag are their blood. They'd be offended to hear you disrespecting them. You're supposed to be an American."

They were completely missing the point, of course, but I was too young to explain it. I can easily do so today: I'm very grateful to live in the country I live in. Even though it's made some grievous mistakes (e.g., slavery, treatment of Native Americans), there is still the promise there that we can be something great. My Busha believed in this. It's why she took her young daughters by the hand and boarded a ship for a country she'd never seen before, leaving behind everything she'd ever known. But Busha never saw a need to be embarrassed by her heritage. She, her daughter (my grandma), and my mother made sure to instill that pride in us. We weren't just Americans, but we were *quintessential* Americans: part of the wave of people who would turn this country into what our history teachers told us was the "great melting pot."

It's a shame my fellow Americans didn't see this. Church was the first place I would experience mistrust and xenophobia, but it wouldn't be the last. No, the worst would come from friends and the general public.

Somehow, probably because at the time I didn't really see the need to keep it a secret, kids in grade school found out I was Polish. We all know kids can be cruel; these kids were.

"Pollack! Pollack! Pollack!" was the chant when I started walking down the aisle in the school bus. What an incredibly offensive, hurtful word. I was a pretty shy, retiring kid in school, but I got into more than one fistfight over this single word. Of course, cussing my tormentors out in Polish only made it worse.

"Hey Mark, how many Pollacks does it take to…?"

"Mark, did you hear the one about the three Pollacks who…?"

Where was this coming from? Elders at church. Kids at school. Even my friends. It seemed that America didn't like immigrants, even though I was born here. I've actually heard "Why don't they speak English?" in church. Wait. I was pretty sure Jesus didn't speak English. I knew that from childhood correspondence Bible courses.

Incredibly, unbelievably, there were people in my church who thought the Bible was written in English. Again, another thing that I cannot make up. People actually believed this.

Later in life, during Holocaust studies at university, I would see a photo, the memory of which still makes me cry. It was taken in a concentration camp. There was a pile of naked, emaciated bodies, stacked to a height of five or six feet. The word "Polak" was written crudely in black across the stack of bodies. A Nazi with a paintbrush delivering one final insult.

It would take decades before I realized that small, insulated communities, with a single ethnicity, and one core religious doctrine, just didn't trust anything or anyone they saw as foreign. The rules were simple. Conform and comply, sing the same tune, and everything will be fine. At least, God be praised, I had the right color skin. My adopted home town was infamous for being all-white, and a 1920s race incident that forcibly removed Black residents is still, embarrassingly, described on history websites.

I instinctively recoiled from this type of thinking. Thankfully, I had some Christian help. My mamaw (reminder, that's Southern for "grandmother") lived through the race riots in Corbin. She would have been in her late teens at the time. When I learned about this very distressing part of our town's history, she told me she'd seen some of it firsthand. Her words still haunt me:

"I was a young girl. There were big crowds of people. They had weapons. They were loading poor Black people onto railroad cars. I knew in my heart Jesus was saying this was wrong. What could I do? I was just a girl. I wanted to say something. It wouldn't have helped."

Mamaw taught me a lot of life lessons, but this was one of her most meaningful. I still remember the look on her face. Sadness. Embarrassment. Resignation. I resolved then and there I would never take that course of action. I wasn't judging her. She is long dead now, but my takeaway was that she was telling me not to make the same mistake. The Alsip side of the family is well known for speaking their minds. In this one case, she felt she couldn't. I'd do my best to make up for it, even if it was in small ways.

A very common word to hear in our small, American, Christian town was the "N word." As soon as I found out what it meant, I found it repulsive. I don't mean in any way to compare my life experiences to others, whose lives I can't possibly have lived and experienced. But being the target of a slur hurts. I started calling people out on this. It made me unpopular.

"Hey, it's just a word," was the excuse I got when I was young.

As I got older and rap music became popular, the excuse changed:

"Well, they're calling themselves that. What's the big deal?"

Maybe as a Caucasian, I should not try to answer this question. Someone much more qualified would be a person who's the actual target of the slur. I would never pretend to truly understand the feelings of another person. What I can do (and have done) is to ask questions and listen to the answers, and try my best to empathize. What I've learned is a very simple lesson. If someone tells me something hurts, then stop hurting them.

At this point, I must confess that within my own family, completely amongst ourselves, the word "Pollack" was used. Given that I would literally throw punches over this word as a child, this admittedly sounds contradictory. Please let me explain.

A rite of passage for any Polish child is when they begin eating solid food and get their first taste of kielbasa (Polish sausage). Not the fake garbage you get at Major League ballparks, but the real stuff, made by a real Polish butcher and filled with all the garlic and incredible spices that make this the most amazing dish on the planet. I'm not exaggerating here: parents in our family would chop the sausage into tiny bite-size pieces, and place them on the tray of Baby's high chair.

Baby would reach out, take a bite, and, almost always, eyebrows would shoot up in delighted surprise and they'd reach out and quickly grab more, as if all the people watching the ritual were going to steal their food.

"That's my little Pollack!" I've heard both grandma and my mom exclaim, as they'd pinch the baby's cheeks.

I don't know if my little story can drive home the difference between an outsider deliberately insulting someone, and someone within the target culture turning the tables, and laughing at the insult—insulting the insult itself by using it in a loving way. There *is* a difference. From my Busha down to my grand nieces, there is a heritage that, if you're on the outside, you probably don't understand.

Because of the ridicule from both my church and the conservative outside world over my heritage, I've adopted a very simple rule in life. If someone tells me that they're offended by what they're being called, I simply don't call them that. It takes so little effort to do this that I don't even think about it. I just don't do it.

It's so incredibly easy to not denigrate another person. My thoughts on this subject are admittedly born from personal experience, but, since I listened so closely in church and read the Bible so many times, I have to admit being influenced by the teachings of another on this subject. I was about to strike out on my own, actually preaching the Bible to others. And I knew one thing:

Jesus would not be the type of person to use racial or ethnic slurs.

We didn't have the Internet or Facebook when I was growing up. When Facebook did come out, I saw a post from a local evangelical congregation. Three men were gathered around an air conditioning unit they were trying to install for the church.

The caption from the church: "How many Pollacks does it take to install an air conditioner?"

People just refuse to learn.

Preacher, Part 1

It might be unfair to say that my church didn't fully appreciate my efforts in learning the Bible and ability to recall scripture at will. True, I did often attend Sunday night Bible study lessons and hit leaders with questions they could only answer with the "you're taking it out of context" canard. But they seemed to look at my overall inquisitiveness, combativeness, and desire to learn and decided it could be molded into something they felt was useful to the church.

And so it was decided that for our congregation's first-ever Youth Day I would be the preacher.

Youth Day was an attempt to involve children more in the workings of the church. Every position, from preacher to communion bearers to song leaders... all were assigned to the children. With careful coaching from our elders, we ran the entire church for the day.

Despite my mounting misgivings about church doctrine, driven by my continued "read the whole thing" Bible project, I was still fully a believer in God and Jesus. And I had an unresolved problem. My mother was still a Catholic. Not properly baptized, according to our church, and worshipping a false "god" (the pope).

I constantly worried about the fate of my mother. On more than one occasion I took church elders aside and expressed my fear that I could never enjoy Heaven if I knew my mother was going to burn in Hell. If you're expecting a compassionate reply here to ease the mind of a worried teenager, then you don't know evangelical churches very well.

Yes, I was told, Mom was destined for Hell, but God had clearly given her a path out—her son was now a knowledgeable Christian, and the Lord was making it obvious that it was on my shoulders to lead her to salvation. And now I had a chance to stand in front of a congregation and preach to her, if only we could get her to attend the service.

No pressure, right?

It should be pointed out early and often that my mother's photo needs to be in the dictionary next to the word "saint." Never has a kinder, more gentle, more caring soul walked the face of this planet. Although I learned many valuable moral lessons from my father, it was my mother who showed, through her actions, the true meaning of acceptance of everyone and forgiveness for any sleight. Still, somehow, my church said she was going to burn, and I was fully indoctrinated. I continued to have nightmares about Mom, but the dreams were expanding to include countless other people who, I realized, were going to be punished.

Yes, even though she was Catholic, she would be in attendance on Youth Day when I delivered my sermon. This was a family event. My parents always turned out to see their children's performances, whether they be ballgames, cheerleading… whatever. My sisters would also be performing church duties on youth day. But, of course, being women, they had subservient roles. In this case they were assigned to the choir. Regardless, my parents were proud of their children and never failed to show up to support us.

My sermon was designed to be generic enough to seemingly apply to anyone, but I made every effort to make sure it spoke to my mother, carefully lacing in "facts" about the one true form of baptism, as it had been taught to me. One of the main things I learned about the Bible was that you can cherry pick it to make it say absolutely anything you want it to. I spent weeks crafting the sermon, pointing out all the errors in accepting anything but the true path to God (hint, hint: you couldn't get there through rogue religions like Catholicism).

My theme was using the Bible as a toolbox. Just like a carpenter would use the tools in his/her box to build a house, God's word contained all the tools we needed to build a healthy relationship with Jesus. But in order for the house to be built, we had to understand the tools and how to use them. I used some not-so-subtle hints regarding my church's teachings and aimed them directly at my Catholic mother: being baptized without being fully immersed was similar to trying to use a wood saw to drive nails. We had the right tool (baptism), but if not properly applied, it wasn't going to accomplish its intended purpose.

I practiced for hours in front of a mirror, imitating the emotional highs and lows I'd learned by watching Revival preachers play to a crowd.

I was ready. I was going to convert Mom.

The big day came. The kids took over the church. Our child choir leader did a respectable job leading the other warbling children through a nice selection of hymns. The communion servers managed to make their rounds without spilling any of the grape juice or wafers (yes, even though Jesus prescribed wine, our church deemed it too sinful and opted for Welch's grape juice). The church youth did a great job with the collection plates, perhaps guilting the faithful adults into larger contributions. Under the gaze of their own children, I suspect that a lot of parents who'd normally slip a few singles into the plate came across with fivers that day. You don't want to let down your kids, right?

Finally, my time to preach arrived. I stepped to the podium and took a deep breath. A flash camera went off from out in the pews. "Oh, hi Mom…" Yes, Mom brought her camera. She later said this was a rare chance to photograph me in something other than jeans and a T-shirt. But I knew in my heart she was proud of me, standing in front of all these people, nominated to be the focus of the whole worship service. I suddenly felt guilty about what I was about to do.

Putting that aside, I launched into my sermon. I carefully laid the foundations of what God expected of us to be true Christians. I had sections of the Bible bookmarked, but, being "blessed" with a good memory, I seldom had to look down at my script. Hey, people were paying attention! I could tell I had the crowd with me.

I fell into the rhythm I'd seen from real preachers… pacing the stage, pausing with my hands on the lectern, and letting silence settle to emphasize my point, raising my voice to an angry pitch when I really needed to hammer the point home. I may have even been guilty of slamming the Bible down at one point with an emphatic "This is the way! This is the only way!"

I don't know if sermons are supposed to get applause, but scattered clapping and "Hallelujahs" mixed with my spoken words. It all seemed to be going so well.

And then the invitation came. The choir took up the inevitable four or five chorus hymn to make sure every fence-sitter in the crowd had a chance to walk up that aisle and be saved.

But Mom never came forward.

I had failed. I failed my mother.

It would be another night when I would cry myself to sleep, convinced I was going to Heaven but Mom was going to burn in Hell.

Preacher, Part 2

Though my own personal reviews of my performance as a preacher had me down as an abject failure, the opinions of the elders of my church were the opposite. I'd apparently demonstrated an amazing knowledge of the Bible and was an effective, compelling speaker. That is, as long as I followed the advice of my church study group and stuck to the parts of the Bible they thought were appropriate. I was encouraged by the church to take this talent further.

Many churches have programs that develop youth for the ministry, and it turns out the evangelical churches in our area were adopting that idea. Our elders learned of a large gathering of potential young preachers to be held at a church deep inside Kentucky Coal Country, and, without really getting a vote, I was elected to attend. Somehow, a few other young men in my congregation were also "elected," so at least I didn't feel alone.

If you haven't visited the heart of Appalachia, it really was (and is) a depressing place. Our group arrived at our destination in a borrowed van late one rainy November evening. I'm withholding the name of the town to protect the innocent. It was a dreary, poverty-ridden place, populated in large part by rundown mobile homes, many with plastic sheets covering broken windows. Yards were littered with junked cars, discarded refrigerators, washing machines, and piles of trash. Half-starved cats and dogs roamed the streets. The decline of the coal industry had hit this town hard, and people were obviously suffering.

And yet, in the midst of all this poverty, there were two modern buildings that stood out due to their relative opulence and wealth: a McDonald's and the local church.

The McDonald's, I could get. Clearly, a rich corporation paid for that building. But the church? The only funding my own congregation had came from member donations. Were the poor people in this town donating what little money they had in order to pay for the upkeep of this building? Some of their own homes literally had no windows, and winter was coming.

Another seed was planted in my garden of doubt. This one would grow large as my journey through religion progressed. Why were people who had next to nothing to give, giving it anyway instead of taking care of their basic needs? Why wasn't the church helping them? The money seemed to be flowing in the wrong direction. I did understand from my own moments of joy at church that people felt it beneficial to attend services. They were clearly donating of their own free will, to build something they felt was important. But it felt wrong.

Inside the church we went, cold rain pouring down on us. A group of young men waited in the pews, displaying various states of nervousness. This wasn't my first rodeo, so I didn't really have a lot of feelings, other than worrying about the people outside this nice warm building. Was the tape holding their plastic window coverings going to survive this dampness?

One notable thing was the audience. Rather than a full church, we would be speaking to a group of old men. We were apparently going to be scored and judged. I admit I've always been a competitive person, so my killer instincts started kicking in. This may have been a Christian competition, but I was going to slay it.

It should go without repeating, but I'll do it anyway—there were no women in the audience. I already knew why, and at the time, accepted the reasoning based on biblical authority. 1

Corinthians 14:34 told me that women should be silent, and wait until they got home before asking questions of their husbands. They were not to speak in church. 1 Timothy 2:12 echoed this sentiment: no woman should be allowed to teach or have authority over a man. Coming from a family of very strong, independent women, these teachings troubled me. In my family, I saw women as equal in all things. Like so many other internal conflicts I had about my church, I bottled it all up inside and asked God to forgive me for my doubts.

But back to the preaching. Honestly, there were some good speakers. Several of the guys who were chosen first in the random drawing made what I knew were biblically sound arguments. Others... well, not so much. As I did during my own preacher's sermons back at home, I sat there silently picking apart the preaching. Anyone who knows any of the descendants of my great-grandfather, who was a traveling, mule-riding preacher that sired twelve children, will tell you that an Alsip was born to debate. It's in our genes. I could never help myself. Everything I ever heard, I questioned. And if there was a contradiction, I wanted to raise my hand and set things straight before proceedings resumed.

When my turn came, I went full Fire and Brimstone on the audience. I hadn't learned just to emulate our regular pastor—I'd attended several revivals at this point. Revival preachers don't so much speak their sermons as they *scream* them at the top of their lungs. I learned how to work myself into a frenzy during certain parts of my sermon. But this wasn't just play-acting—I actually felt something as I spoke. I was convinced in that moment that God was speaking through me. I abandoned the reasoned appeal from my sermon to my mom and dove into a horror-based, book-of-Revelation-inspired tirade based on all the fears I'd been holding inside since watching the Whore of Babylon movie described in an earlier chapter.

The crowd seemed to be eating it up. The old men were nodding in agreement as I hammered my points home, sometimes reaching out and touching their neighbors on the shoulder and pointing at me on the lectern. Reading lips, I distinctly saw the word "talent" mouthed.

The rain poured down harder outside. Thunder clapped, interrupting my sermon. I paused to think about the people huddled in the mobile homes. My train of thought derailed. As I resumed preaching, I felt I was delivering the wrong sermon. I should have adopted a theme of an aggressive Jesus going after the moneychangers, and demanding answers on why we were warm and comfortable in a brick building with solid windows while just outside, people struggled to keep broken windows covered against the downpour.

As we left the church that night, our youth coordinator slapped me on the back and congratulated me on the best sermon of the night. I looked over at a mobile home across the street, stray cats huddled in an old clothes dryer lying in the yard, and hoped that the inhabitants of the trailer had guaranteed themselves a place in Heaven by helping build this nice warm church.

Because they were certainly living in Hell right now.

Adult Education

Katherine (name changed for obvious reasons) was the prettiest girl in our church. When I watched her walk through the building, it was like time slowed down and everything moved in slow motion. Wisps of her hair would float in the breeze, framing a perfect face with flawless white teeth that were displayed in a perpetual smile. I was fourteen or fifteen. She was a few years older. I didn't realize it at the time, but I had my first real crush.

She was a decent piano player and would sometimes fill in for the elderly lady who normally accompanied our hymns. And it was on one of her substitute days, on a beautiful spring morning with the church windows open to capture the breeze, that Katherine stepped in front of one of those bright windows to sit down at the piano.

She was wearing a light pink dress. Sunlight was streaming through the glass, backlighting her. A breeze blew the dress against her body.

Wow.

I was a very sheltered child growing up. What was *this*? The shape I saw silhouetted in front of me was... well, it was mesmerizing. There were curves and bumps I'd never really paid attention to before and... well... I started having strange feelings I didn't quite understand. I shifted uncomfortably in the pew. Something strange was happening *down there* and I didn't want anyone to notice. I was embarrassed, but still spellbound by the vision of Katherine. My dad elbowed me sharply to stop my wiggling. The hymn was beginning.

Sex education was a taboo subject when I was a youngster in the South. The only way I'd heard of it was a pastor ranting about "things that shouldn't be taught in schools." My parents never had "the talk" with me. I was left to go it alone. Sure, guys on the playground would talk about girls, but the amount of realistic information to be found in such conversations could fit on a pinhead, with plenty of room left over for angels to dance on top.

All that church taught me was that attraction to the opposite sex was bad. Well, bad until I got married. I'm glad they clarified that because at first, I was wondering if they were saying I should be attracted to the same sex. Don't worry, they finally covered that subject too—and as you can imagine from today's religious/political environment, the views on that topic were not very kind. Another one-way ticket to Hell, as I recall. Guys were a definite no-go. Girls? Wait until your wedding night. Everything you needed to know, you'd figure it out after the wedding cake was finished.

And so, it wasn't until my freshman year of high school that I finally encountered sex education. Even in conservative Kentucky, it was finally decided that the topic should be broached in the biology classroom.

It was a farce from the get-go.

First, students had to have a parental permission slip to be in the classroom that day. I think my dad signed with a sigh of relief. He was a kind but shy man. Talking about sex just wasn't his thing. Why not let the school handle it? (The latter, in my humble opinion, is not such a bad idea. Why not allow kids to learn from certified teachers?)

The problem is, in my religious, conservative state, at least half of the kids had parents who opted their children out. I remember one in particular, a friend of mine. I'll call her Amanda for lack of a better false name. Before class began, the teacher asked those who didn't have approval to please leave the room. Amanda was among the many who walked out.

I really didn't learn much from the class, to be honest. Note to adults: kids and teens can tell when you're embarrassed and would really rather be somewhere else, talking about anything other than what you're talking about. A red-faced teacher rolled down a couple of charts, one of a male body, one of a female body. He took out a wooden rod and haltingly pointed to what were obviously different parts on the two sexes, and gave them names that my parents told me to never repeat again when I said them at home.

"Penis."

"Vagina."

"Mark, go to your room!"

Back in the classroom, the teacher was quickly concluding the sex education lesson. "And so, men and women, having these different body parts that God gave them for the purpose of marriage, get together *after they're married,* and that's where babies come from." The end.

Nothing about birth control. No pills, no condoms. Not a single word about sexually transmitted diseases. Seeing images of penises and vaginas wasn't of much help if you weren't told what to do with them (or the dangers of not using them with protection). I know this will seem incredible, possibly unbelievable, to modern-day readers, but it really was possible to keep children in such a sheltered state in that day and age. The Internet was still over a decade away. Sex was an evil thing; a tool of the devil. At the end of the so-called lesson, not a single kid in that classroom had any more clue about what caused babies and how to prevent it than they did when they walked in.

Speaking of walking in, the door opened and all the students who'd left were ushered back in. Amanda sat down at her desk next to me and asked what she'd missed. "Nothing, really," I replied.

That very same year, Amanda dropped out of school, pregnant.

Another teenage girl in poverty-ridden rural Kentucky with a baby to care for. The conservatives at my church derisively referred to these children as "welfare babies" and ridiculed the girls for their loose morals and not obeying God.

"She should've kept her legs closed," said the guys in gym class. These were the same boys who would often brag about trying to get her to spread her legs in the first place.

This pattern would repeat itself throughout my four years of high school. One year we lost a cheerleader due to "medical reasons." Oddly enough, her boyfriend, the presumed father of her child, was a star player on the football team and remained in class. He went on to lead the school to a state championship.

Too bad Amanda or our cheerleader never got the chance to enjoy such an opportunity. They were too busy at home, changing diapers.

Spilling Your Seed

This is a story I tell with more than a moderate amount of embarrassment. Perhaps this comes from a former evangelical's sense of modesty and lack of sex education, when we were taught to be ashamed of topics like this. But as with my other stories, I promise there's a modern-day lesson here. So, I'm going to plunge right in. You can't tell if my face is red or not right now, correct?

Around the time that I noticed something... different... about Katherine that day in church, particularly the way her body was shaped, an embarrassing problem reared its ugly head. Some nights, in the middle of particularly vivid dreams, I'd wake up with a wetness in my shorts. I was horrified. Was I wetting the bed?

I recalled a verse I'd read (and obviously misunderstood) from my first reading of Deuteronomy:

"If there be among you any man, that is not clean by reason of uncleanness that chanceth him by the night, then he shall go abroad out of the camp, he shall not come within the camp."

"Uncleanness that chanceth in the night?" In my childhood innocence, and because, like today, evangelicals are fighting like mad to keep sex education out of schools, I simply thought that some poor Israelites were wetting the bed. Of course, as it turns out, it wasn't *that* kind of wetting they were talking about in the Bible. But nobody was about to teach us about wet dreams. Not in church, not in school.

One thing I'd come to eventually learn about evangelical religion is that it was hell-bent on making sure we were ashamed of our bodies. And often it would do this without children having enough education to understand what they were to be ashamed of. I'm not claiming that church needs to be the place where sex education takes place. Rather, looking back from my early childhood to today's America, it's often religion that attempts to prevent children from being educated *at all*.

Anyway, back to the dreams. Here I was, a few years later, and I was waking up with wet spots and the tingling feeling that came along with them. And to be honest, the tingling was a rather enjoyable feeling (though I didn't know exactly what it was at the time). With absolutely no help from any adult in my life, despite our conservative state's prohibition on sex education in the classroom, I put two and two together: Deuteronomy wasn't talking about bed-wetting.

Forearmed with this knowledge, I revisited some verses in Leviticus and found out that these types of "emissions" were unholy and made me unclean. I was deeply ashamed, kept it quiet, and pushed it to the back of my mind. But then one night in a church revival I learned that this was a sin—possibly a deadly one.

If you've never attended an evangelical revival, you've been missing an interesting show. Revival week at church was when all the visiting fire and brimstone preachers came calling. Most of the sermons were delivered with red-faced shouting, lectern banging, and an angry man strutting across the stage waving a Bible at us.

One particular sermon was on the cost of disobedience to God. For reasons that I'll never know or understand, the guest pastor that night chose to tell the story of a man named Onan, from the book of Genesis.

It seemed that Er, the firstborn son of Judah, performed some unspecified act to displease God. So God struck Er dead on the spot. Now, there was a problem. Judah wasn't going to get grandchildren out of his firstborn son's line. For a patriarchal culture, this just wouldn't do. So, Judah ordered another son, Onan, to sleep with Er's wife and conceive a child.

Onan wasn't really happy with this idea. It wouldn't be his own child. It wasn't his wife. Moral and ethical implications in mind, whenever he had sex with his brother's wife, Onan withdrew early and "spilled his seed on the ground."

God's reaction, of course, was to strike Onan dead. God seemed to strike a lot of people dead in the Bible.

I was horrified. I had been guilty of seed spilling. Was God going to punish me too? The answer would turn out to be yes. But, oddly enough, this time my conservative evangelical church wasn't the source of my information.

Being baptized Catholic and still making frequent visits to Milwaukee, I had the chance to attend Mass with my Wisconsin family. Most of the service was in Latin, but, either by coincidence or an act of God, one Sunday a priest stood up and did a sermon, in English, on the sins of contraception and masturbation. For biblical support, he went right back to the story of Onan spilling his seed. Yes, this was indeed sinful. We were, according to the priest, wasting what he referred to as "the seed of life," and we were doing so in a way that God hadn't intended. We were doing it for our own pleasure, not to procreate, as the Lord had planned.

Thank God for big city churches. They're placed, obviously, among large, presumably better-educated populations, where there are fewer chances (presumably) for sex education to be kept out of schools. Nobody in this particular Catholic church in

Milwaukee was afraid to put specific names on things. Contraception and masturbation? I needed to look that up. I often speak of the two sets of encyclopedias my parents kept for their children. There was actually a third book, a medical encyclopedia. When we returned to Kentucky after vacation, I grabbed that volume and went off to my room for a little private education.

Even though the encyclopedia laid things out clearly and assured me that all of this was natural for humans, it was still competing against a book that I had sworn to hold as the higher authority: the Bible. I constantly lived under the threat, delivered by my church, to ignore any and all information that contradicted church doctrine. As was the pattern throughout my youth and early adulthood, I was torn between two worlds: science and religion.

Flashing forward to modern-day America, it's alarming to see incidents of book banning and prohibiting certain topics from being taught in school. This religiously-driven movement does a great disservice to our children's education. How do we know we're sinning when they won't even describe the sin to us?

I still think of my friend Amanda, pulled out of sex-ed class, becoming a mother before she was even old enough to drive a car.

Slavery

Though my primary focus in school was on science education, and many of the conflicts in my life came from the inevitable disagreements with fundamentalism, there were numerous significant conflicts involving equally important topics, such as history and morality. As would become apparent in my adult years, when Florida schools were forced to teach that slavery benefitted Africans (by teaching them a trade), fundamentalism wasn't solely targeting science.

Throughout all of my Bible study, I couldn't avoid the references to slavery. Exodus gave instructions on beating your slaves. In Leviticus, God gave the Hebrews permission to take people from other nations and keep them as property, handing them down from generation to generation. The New Testament didn't make things any better. Jesus said in Matthew that he hadn't come to overturn any of the old laws. The fact that I was a member of a religion that worshipped a being who'd ever condoned slavery in the first place gave me constant nagging doubts.

Fundamentalist Christian explanations could never quite overcome the problem presented by the dozen or so verses in scripture that gave the green light to owning other human beings. Some people told me that Biblical slavery was just indentured servitude—people offering themselves into service until a debt was paid off. No. Leviticus clearly says that was not the case.

Others tried to tell me that this was "just the custom of the time, and the Bible is trying to be historically accurate." My thoughts in response to this were that God not only took the time to tell us not to kill and commit adultery, he also went to the trouble of telling us we couldn't eat shellfish or pork, and couldn't mix different types of cloth in our garments. Why was there no eleventh commandment, forbidding the practice of owning humans?

This was, and always will be, one of the wedges that irrevocably drove me away from religion. Nobody in theology has ever been able to adequately explain away or offer enough apologetics for slavery. Incredibly, our church actually taught that Africans had, at least in part, been done a favor. If not for their captivity, they might not ever have had the chance to hear the Gospel. Their masters gave them the chance to learn about Jesus. This was a literal teaching of my church.

Kentucky was a border state during the Civil War and never officially outlawed "that peculiar institution." When I was growing up, forty miles north of the Mason-Dixon line, the war between the states could have just as easily been fought a few decades before, rather than in the 1860s. Sympathies and apologetics for "state's rights" ran deep in my small town. Never mind that the rights in question were to possess others as property.

And so it was that one day I sat in an eighth-grade history classroom and got a watered-down education on slavery—teachings that are alarmingly working their way back into public schools today.

"Slavery was not necessarily a good thing," my history teacher intoned, "but it was nothing like you hear some people talking about today. I want you all to think about this: a good slave could cost several thousands of dollars. Who on Earth would pay that much money for something and then treat it badly?"

She continued, "slaves were important to the well-being of their masters. It benefitted the owners to treat their property as well as possible. The propaganda you hear today about mistreatment of Africans cannot be true. It would have been economic suicide to treat people this way."

This was an American school in the Bible Belt in the late 1970s. How far have we progressed since then?

I wanted to fight back. I wanted to raise my hand and say something. Did anyone consider freeing the slaves and paying them an honest wage? Treating them as equals? But I lived in a small southern town where there were literally no Black people. I knew the beating I'd get on the playground if I uttered a single word in objection to what I'd heard in class. The "N word" was used quite liberally in Kentucky schools—I first heard it in the 3rd grade at Oak Grove Elementary. I asked what it meant then, was told, and simply went on with recess. There was nobody around to tell me it was wrong. I heard the word repeated by church members. When I hear it today, I recognize my younger self, without the benefit of a more enlightened education.

Our church's teachings on race relations? Not good. I heard more than one person in our congregation say that they didn't believe God intended for the races to mix. They never really offered any Biblical justification other than obscure conspiracies about the cursed line of Ham, one of Noah's children. It all occurred because Noah's sons found him naked and drunk. Noah actually cursed Ham's son, but "curse of Ham" stuck. According to some evangelical traditions, the descendants of Ham migrated to Africa and became a cursed race, marked by the color of their dark skin. You can't make this stuff up. It was taught by my congregation, and I've heard White Nationalist fundamentalists repeat it.

Racist teachings weren't peculiar to my own congregation. It wasn't until the late 1970s that the Mormon church officially renounced their long-held position that Black people were suffering under a curse from God. Their reasoning? The curse of Ham.

I look back now on my own history class and can feel the injustice inflicted on students and teachers in the state of Florida. It's now required in that state to teach that slavery put the enslaved onto a valuable career path.

I'm not sure how owning a human being is necessary to teach them a trade. In fact, skills like metalworking were already known in West Africa when the slave traders came calling. No lessons needed. Other knowledge, like rice farming, was completely foreign to American slave owners in states like South Carolina, who prized African captives who were already experts in growing that crop.

And yet, from my school days up until today, we have public schools trying to whitewash one of the most horrible things our country has ever done.

We've been removing our history from schools since I was a teen, and it seems to be getting worse. Home schooling, which has little if any government oversight, can skip the topic of slavery entirely. I recently encountered a home-schooled fundamentalist White Southern Christian who claimed that the South only held a few slaves; that the Union troops made up the whole American Slavery story as a propaganda campaign against the Confederacy. In this telling, Confederate troops joined together with their Union comrades to free the few slaves that were being held.

I couldn't respond. I just turned and walked away from this woman.

But not before she informed me that she was in charge of the Juneteenth celebration at her church... would I like to come?

Guns and Bibles

Members of my church rejoiced when the Bacon Creek Gun Shop burned to the ground. All of the gun purchase records, registrations, evidence of ownership… everything was suddenly gone. It would take years for me to truly understand why my congregation felt this way.

America wasn't as gun-crazy when I was young. True, in Southeastern Kentucky, guns and hunting were a way of life. I fired a gun for the first time when I was twelve years old. It was a classic L.C. Smith double barrel shotgun, originally owned by my papaw (Southern for "grandfather"), who spent his life in the coal mines before passing away before I was ever born.

Of course, I landed flat on my butt from the recoil of my first shot, but I couldn't help belly-laughing along with my dad, my shooting coach, who was helping me hold the gun up. It was a rite of passage in my family. Dad landed on his butt as a kid too. But he turned into an expert marksman in the Army, and taught me many valuable lessons about firearm ownership; none of which are on exhibit in 2024 America.

Before the National Forest Service saw fit to cut down all the hardwood in our section of the Daniel Boone National Forest, it was a wonderland. My father started training me young, as his father did with him: long walks along ridges and game paths in a wilderness where even the second or third growth oak, hickory, and walnut soared so far overhead it was hard for me to see the tops. It was a paradise for a child who loved nature. I probably spent as much time studying the biology of trees as I did hunting. Then logging companies came in, poor Appalachian workers with no income were lured into chopping down all the hardwood, and the forests became a wasteland of scrub brush. The lumber companies moved on, and the workers went back to their lives of poverty.

Our early forays into the forest were all training runs. I didn't carry a gun. I was there to learn to walk quietly, to stalk game, to learn how to navigate. Some of my fondest memories of my father are those long treks in the woods, halting suddenly when he raised a fist, then pointing to a tree where small pieces of shells were being dropped by the squirrels up above. I may have been training to become a hunter, but I also developed a deep appreciation for nature. It's amazing how much you miss when you don't slow down and look. Ironically, this training would be of great value later in life when I would trade my gun for a camera and become a nature photographer. If you can't remain invisible to animals, you're going to come home with a lot of empty rolls of film (well, today, it would be empty memory cards).

To those who find hunting to be cruel: I've always looked down on big game hunters. Shooting an animal to put it on display is an unthinkable act of cruelty. To my father, killing something you did not intend to eat was akin to sin. In fact, growing up in an actual shack with seven siblings just after the depression, his family often relied on hunting to feed themselves. There was no local supermarket. If there had been, Dad's family didn't have the money to buy food.

Hunting for food, though? I never had a problem with that. Both a cousin and an uncle were professional butchers. I had no illusions where the meat on our table came from. Besides, according to our church, all the animals and the rest of nature were all put there for our exclusive use (Genesis 1:26).

Hunting was a popular pastime for many members of our church as well. After-sermon Sunday discussions outside the church door always featured animated conversations with the onset of deer, turkey, rabbit, or squirrel season. But I was confused over the joy over the town's only gun shop burning to the ground.

However, listening more closely to the conversations, I heard words such as "Now it'll be harder for the government to find our guns and Bibles when they come for us."

What?

As time went on, and I, the ubiquitous question-asker, quizzed my fellow Christians, it became clear something more dark and sinister was going on. These people weren't just preparing for deer season. They were preparing for the End Times. The book of Revelation, which I was now intimately familiar with, came up quite often in conversations.

Looking back on my religious education—nay, *indoctrination*—it is clear that I was taught to expect to be constantly persecuted for my Christian beliefs. It's a core belief of fundamental evangelicalism. I didn't need my church to warn me of this. I found countless examples in my daily Bible readings. For example:

John 15:19–20
"If you were of the world, the world would love you as its own; but because you are not of the world, but I chose you out of the world, therefore the world hates you. Remember the word that I said to you: 'A servant is not greater than his master.' If they persecuted me, they will also persecute you. If they keep my word, they will also keep yours." (Emphasis added).

And so, church member after church member told me how happy they were that the gun shop was now a pile of ashes. The place could be rebuilt. Nobody was hurt by all the ammo cooking off. Nothing was lost; much was gained.

There were no more records of gun ownership.

When the government came to confiscate my congregation's Bibles, the true believers were locked and loaded.

This story may seem a diversion from the plot line of this book, but I believe it's fitting. Recall that part of my mission in writing these words is to explain, from an insider's point of view, the underlying "logic" behind the right-wing religious extremism we see in America today. Incredibly, I have religious relatives who actually do hide their guns and Bibles. They're absolutely convinced that some mythical world government is going to come for them, and they're prepared to fight back.

Since my day, the weapons have grown more powerful, easier to obtain, and greater in number. The evangelicals who possess these weapons seem to have grown louder, and they're growing in number as well. A popular bumper sticker here in the Bible Belt reads: "I believe in 1 God, 2 genders, and 100 reasons to own a gun." Guns and religion are becoming more intertwined.

I miss the early days of hunting, when it was just me, alone, deep in a forest, holding the same instrument my papaw held when he was young. I could still see the spots on the metal where his fingers had rubbed off the bluing. It was my only connection with him.

When I learned to shoot and hunt, I had no idea there were fellow church members who honestly, deeply expected me to participate in an End Times battle. This prospect frightened me greatly. I recalled the lessons and fears described in the "Whore of Babylon" chapter. Was my religion preparing me for armed conflict?

The Crying Indian

My church taught me early and often that God gave humans dominion over the entire planet, including its resources and animals (we weren't allowed to refer to humans as "animals," even though we technically are—God made us more advanced, the church said). When alarm bells started ringing in the 1970s about the harm that pollution was doing to Earth, nobody in my congregation took it seriously.

Once again, the conflict between verifiable science and fundamentalism would come into play.

"Humans can't do anything to affect the Earth. It's too big. God wouldn't let anything happen to it. He created it. Stop worrying. He knows what he's doing."

"It's our planet, and we can do what we want with it. The coal was put here for us. The oil was put here for us. The water was put here for us. God said it, I believe it, and that's the end of it."

A famous commercial came out at that time. It's known, perhaps somewhat culturally insensitively, as "The Crying Indian." "Crying Native American" would have been a better choice, but this was the 1970s, and people weren't really much into considering how inappropriate labels could do harm.

In the commercial, a Native American (played by an Italian-born actor, because apparently Hollywood didn't think Native Americans could act?) paddles a canoe down a polluted river. Factories belching dirty smoke into the air are seen in the background. The canoe lands on a shoreline covered in litter. The announcer intones "some people have a deep abiding respect for the natural beauty that was once this country." As the actor walks to the side of a road, a car drives by and throws a bag of trash at his feet. "And some people don't," the announcer concludes.

The "Indian" cries.

The commercial hit home for me. Kentucky was a beautiful state, but it was common for people to throw fast food rubbish out of their car windows as they drove down the road. Roadside dumps with mattresses, washing machines, and every imaginable form of garbage littered the sides of forest trails along which we hunted. Coal mining was destroying water sources. When my buddies and I hiked our forests, we were puzzled by all the orange streams. (I'd later learn this was due to the formation of sulfuric acid and iron from carelessly dumped coal tailings.)

I was upset that the Native American cried. I wanted to cry too, sometimes.

Every year, our family took a trip back to my birthplace, Milwaukee, Wisconsin. Along the way, we had to drive by the city of Gary, Indiana. Anyone who ever went within miles of that city knows it was Hell on Earth. On certain days, the pollution in the air could make your eyes burn and water from miles away.

We really didn't seem to have a good plan for taking care of the planet God had given us. My parents and grandparents told stories of drinking from natural springs and swimming in lakes and rivers when they were young. You couldn't do that when I was a kid. Everything was too polluted. And few seemed to care.

Our schools cared. We were being taught about the dangers of toxic chemicals. Scientists were beginning to talk about damage to something called the "ozone." I looked it up in the encyclopedias at home. This sounded dangerous.

It's interesting to go back to the history of the '70s and '80s and read that fossil fuel companies knew even then that their emissions were having a negative impact. And Kentucky's coal companies? Most mined the coal they could easily get to and then abandoned their holdings (and miners like my papaw) leaving it to future generations to clean up the mess. The few remaining companies are staunchly supported by politicians and citizens who make no secret of their conservative, religious leanings.

My fundamentalist religion taught me that the planet was ours to do with as we saw fit. My secular education told me differently. I could visit the creeks and springs where my ancestors once swam and obtained drinking water, but I couldn't play or drink. It was too dangerous.

In 2024, when I look at pollution and climate change denial, I can trace the phenomenon, at least in part, to my fundamentalist upbringing. God gave us this planet. Apparently, if we want to trash it, that's our prerogative.

Religion in School

Despite religion (or at least prayer) being formally removed from public schools by the Supreme Court before I was ever born, it never really went away. We began class every day at Oak Grove Elementary by reading our teacher's selection of Bible verses. We stood and pledged allegiance to a flag, with the words "under God" having been added in 1954 due to pressure being put on President Eisenhower by religious groups and pious politicians.

(Eisenhower would also be responsible for forcing "In God we trust" onto American currency, circa 1955-1957, after enacting a law that made it the official motto of the country.) In my grade school years of the 1960s, people would point to our pledge and currency as evidence that we were a "Christian country," not realizing this was a new, forced, standard. Sixty years later, modern-day evangelicals will swear that the pledge and the currency have always been that way.

Everyone in my grade school class clamored to be the honored one chosen to do the daily Bible reading. Religious indoctrination would continue at least until the year before I began high school. In the eighth grade, under the guise of "history lessons," Christian missionaries would attend Whitley County Middle School and tell Bible stories. One husband and wife pair will always stick in my mind. They used a type of felt easel with Velcro-backed cloth Bible figures and told stories by moving the characters around on the easel. Very boring, yes, but portable computers and digital projection screens hadn't been invented yet, so one made due with what one had.

As with many religious authorities in my life, I'm not sure these particular missionaries were very well educated on their background material, or perhaps they didn't really understand the target audience. Their most memorable lesson, at least to me, was the story of Elisha and the Two Bears from the book of 2 Kings. As the story goes, Elisha was out on a stroll when some young children started making fun of his bald head. Elisha cursed the children, so God sent two bears out of the woods, and they ripped the children to shreds.

All forty-two of them.

I was taught by my parents to always show respect to my elders and would have never made fun of anyone for lacking hair. I would have been punished if I did. But the death penalty seemed a bit extreme. To be honest, through all my many readings of the Bible, the story of Elisha didn't really take hold in my mind alongside the flood, the killing of the firstborn of Egypt, the sacking and slaughter of all the inhabitants of many cities, and all the other large scale horrors... but seeing the bears chase the children around the felt board really brought the story to life.

Some of the cloth kids even had detachable body parts to graphically drive the point home. A little cloth arm fell off a dismembered child and hit the floor. Nobody noticed.

I don't know that this lesson really had the effect the missionaries had hoped. The end game seemed to be to reinforce the teaching that God was going to punish evildoers. Maybe some of the class of eighth graders took it that way. But to me, a biblically-literate young adult, it planted another seed in my garden of doubt. I left class that day with yet another example of God inexplicably killing children who were either completely innocent, or at worse deserved a good talking-to from their parents. I harkened back to my father's firm, gentle, effective punishment when I was caught stealing as a youngster. How differently that would have gone had he sicced a wild animal on me!

Forty-two youngsters torn apart by bears because Rogaine hadn't been invented yet.

Cosmos

In 1980, the seventeenth year of my life on Earth, three significant events would occur. The first was the release of Carl Sagan's series *Cosmos* on public television. I'd been in love with astronomy since my parents gifted me with my first telescope five or six years earlier, but living in a small town long before the advent of the World Wide Web, I was cut off from discussion of the science and discoveries surrounding the science of astronomy. All I had were dated library books and monthly issues of a few astronomy magazines.

I didn't really have an influential science show to latch on to as I grew up. Mr. Wizard ended before I was born. The fabulous Bill Nye was still decades away.

Cosmos changed all of that.

I happened upon the television series completely by accident. In my day, there was no cable or satellite television. There were no streaming services. When radio propagation allowed, we received all four existing networks (ABC, CBS, NBC, and PBS) via an antenna pointed toward the nearest large city, Lexington, about 80 miles away.

PBS (the Public Broadcasting System) played a large part in the education of the Alsip children. It's where we cut our teeth on Sesame Street, the Electric Company, and other educational shows. I would never learn to paint, but I was transfixed watching Bob Ross paint his happy little trees every afternoon when I got home from school. There was just something hypnotic about the man. You couldn't watch him and not feel good. He was a gentle soul who loved art, and shared that love in a way that was infectious and easy to understand. Unbeknownst to me, Ross was teaching me elements of composition, something that I would put to good use when I eventually took up photography and saw my work published.

Those sneaky educational shows.

It was after the closing credits rolled at the end of another glorious Ross episode that a preview appeared for a new series. It was named *Cosmos*, and it was about astronomy! This was something I absolutely had to see. If memory serves, it was played in Kentucky rather late at night. Our parents had fairly strict bedtimes for my sisters and me, especially when school was in session, but as I approached my eighteenth birthday, the rules were relaxed. Mom and Dad knew I was growing up. On Fridays and Saturdays, I was allowed to set my own bedtimes, provided that I was up in time to do my weekend chores (like mowing our one-acre lot with a push mower, or washing cars—Dad never did forget about that rainbow lesson from chapter one).

Late night television was already opening my eyes to a new world. My sense of humor would be indelibly shaped by Monty Python's Flying Circus and the Benny Hill Show. Wolfman Jack's Midnight Special introduced me to rock groups that would influence my music tastes for decades.

But then, above all, there was *Cosmos*.

The premier episode started off in a rather understated manner. One man standing alone on the top of a rise overlooking an ocean. He started out talking about the Earth, not space. But he had such an intriguing way of speaking, a master story-teller's command of the screen, that I was instantly hooked.

The most significant episode of the show, to me, was actually the second, where Sagan described in less than sixty minutes more about evolution than I had been taught in my entire life up to that point. Recall that I lived in the conservative South, where evolution, if it was given any classroom time at all, was nothing more than talk about an old man with a grey beard who noticed that some finches and turtles were different on neighboring islands. Religious pressure kept evolution out of schools. It was an outright sinful subject in churches, and railed against on a regular basis by preachers.

Without ever once mentioning religion, without ever presenting evolution as a controversial or religious topic, Sagan laid out all the science in a way that seared the subject into my mind. Finally, things began to make sense. The similarities between me and the gorilla I'd seen at the Milwaukee Zoo all those years ago? Now I had a natural explanation. All of the straw man arguments I'd heard at church ("they say humans descended from monkeys") were swept aside. These weren't our ancestors. They were our cousins. I didn't see a conflict with religion. Carl wasn't teaching theology. He was laying out facts that I could observe for myself.

Why were my church and school hiding this information from me?

It would be another year before I entered university and was able to build on some of the information being presented. I took notes in my head. DNA. I knew I needed to remember that. The fossil record? My church elders were telling me God planted fossils to tempt me. I'd already discarded that teaching, despite the warnings I'd burn for it. The Bible told me God didn't tempt us. Not only was my church contradicting science, it was contradicting the Bible itself.

The constant inconsistencies injected even more conflict and trauma into my life. I became unsure about everything. My self-esteem, my belief in myself, hit rock bottom. It was difficult to be sure of anything. My congregation told me that what I was learning in secular school was all wrong, yet I was finding flaws in my religious teachings as well. I was being asked to make choices where none of the outcomes made sense.

Nothing about *Cosmos* ever presented an "either/or" choice between church and religion. The series was merely laying out information that I could research myself. As kids, we found fossils all the time in road cuts. But only fossils of sea creatures. At church, they told us this was because of Noah's Flood. But the flood was supposed to have drowned *all* life on Earth. Why wasn't I finding human, dinosaur, and other animal fossils? (The answer, I would find out later in college, was that modern-day Kentucky spent a great deal of Earth's history covered by a shallow sea. There were no land-dwelling animals here to be drowned.) Somebody hadn't been telling me the whole story. Again, I resolved to learn more. Looking back at my past, recalling the information I was picking up from one amazing television series, I would deliver a warning to modern-day Young Earth Creationists: kids aren't foolish. You can try to hide the truth from them, but they're going to figure you out.

Unless, of course, the early indoctrination takes hold, and the child never escapes. Removing science education from schools and/or removing children from public schools altogether, placing them in private or charter schools with little oversight, would be an effective way to prevent science from reaching our children. But for the grace of *Cosmos*, there could have gone I.

I almost wept as Sagan described the loss of the great library of Alexandria. All those books, all that knowledge, lost forever. I loved my local library. I practically lived there. How could anyone neglect or destroy a library? Other episodes of *Cosmos* spoke again of books being destroyed or banned, this time by churches. How could anyone ban a book? What could possibly be so frightening about words that a religion would rise up in opposition?

Thinking it over, I realized there were no books on evolution to be found in either our school or public library. I was essentially living under a book ban. It did not feel good.

Sagan made a bold but precise prediction: it was expected that eventually planets would be discovered in star systems other than our own. This had been a dream of mine since age thirteen, when I first read Isaac Asimov's *Foundation Trilogy*. I was ridiculed for bringing up the topic in church. God created life in one place, and one place only. Now, here was a scientist explaining why this was likely. He did warn that for the duration of *Cosmos*, he would be speculating, and he would be very careful to clearly separate speculation from fact. This humble honesty impressed me, and I'd incorporate it into my own life. But the growing evidence that there could (should?) be other planets out there was intriguing.

It wouldn't be much more than a decade before the first extra-solar planet was announced. Then more discoveries followed. Evidence of other worlds trickled in by the handful at first, then it turned into a torrent. As I write this, the number of known planets is in the thousands, and it seems like it's the rule, rather than the exception, that other stars host planets. Science was *predictive*. Not predictive as in a horoscope that made vague prophecies about my love life—specifically predictive as in "we know how stars and planets form and, when the technology becomes available, we expect to find planets all around us."

Sagan went on to explore the difference between simply believing in something and having evidence that that "something" was true. There were examples of many scientists who were so obsessed with their pre-conceived notions that they could not abandon them even in the face of overwhelming evidence. Ptolemy drew up elaborate models to "prove" that the Earth was at the center of the cosmos—and he was wrong. Perciville Lowell was convinced he saw canals, evidence of life, on Mars—and he was wrong.

I learned a great truth here: it's okay to admit when you're wrong. It's actually a fundamental pillar of science. Young earth creationists were always pointing out to me how often science "changed its mind" (for example, when something new was discovered concerning evolution), and so therefore science couldn't be trusted. I discovered that I should be more suspicious of those who could *not* change their minds in the face of evidence.

If you don't change your mind when you're proven wrong, you're doomed to live a mistake-riddled life (spoiler alert: in just three short chapters we'll see, via the explosion of a moonshine still, just how important it is to learn from your mistakes. Wait for it…).

I thought of my fundamentalist church elders who insisted the Earth was only 6,000 years old, despite overwhelming evidence to the contrary. They would not change their minds. I wondered how they would feel about a science-based doctor who'd initially diagnosed them with arthritis, only to later find out it was cancer. Would these elders want the doctor to keep insisting on arthritis, or admit they had it wrong and start treating the cancer?

I mentioned three significant life events at age seventeen. It was in this year that I'd just received an upgrade to my amateur ("ham") radio license. After a rigorous FCC exam, I was granted an Extra Class license, giving me access to all but a tiny sliver of the radio spectrum available to "hams." Sagan touched on receiving radio signals from outer space in several episodes of *Cosmos*. In an epiphany, I realized that my new radio privileges would let me listen in on the universe.

I was in love with the planet Jupiter. The Voyager probes had just done their flybys of the planet, revealing detail I could never hope to see in my small telescope. After reviewing the setup I'd need, I put everything together one night and sat at my operating station, whose window just happened to face to the east, and waited for Jupiter to appear over the horizon. I was greeted with the sounds of what sounded like waves crashing on a beach— exactly as scientists had predicted. I was listening to radio waves generated by that bright planet in the sky. The signals had left Jupiter approximately forty minutes earlier, just as had the planet's light that was just now entering my eyes.

On some nights, after "tuning in" to our largest planet, I'd take my telescope out into the back yard and watch its moons in their slow, steady dance around the gas giant. To imagine that the Church condemned Galileo and sentenced him to house arrest for simply describing what I could see with my own eyes! Would it be arrogant of me to say I felt a sort of kinship with scientists of the past? No, of course, I wasn't being burned alive or arrested for my scientific views, but my fundamentalist congregation certainly reminded me that I *would* burn eternally if continued down the scientific path.

I'd learned another important thing about science, thanks to my observations. I could recreate—I could repeat—experiments done by others, and get the same results. I didn't have to take somebody's word for it. The contrast with fundamentalist religion was obvious, but I dare not bring it up during Sunday meetings.

I would be remiss to not mention at this point the positive influence of yet another devout religious person on my life. My amateur radio mentor. In small towns such as mine, there were no formal electronics classes I could take. Self-learning was the only way. Except that my father, on his route as a mailman, happened to have befriended a kind Catholic man who, along with his wife, ran what could almost be called a center for kids in their home. They were loving grandparents, but they adored all children and seemed to have adopted half the town.

The doors were always open at their home (but please always call first—they did stay busy). Lessons in photography, cooking, electronics, even just a quick chat about how life was going... there seemed to be a steady stream of youth traipsing through the doors of what would become another set of grandparents to me. One of the many good acts these good Christians performed was setting up an amateur radio service so that the children from the local Catholic boarding school could speak with their families back home in Central and South America.

My mentor once called me at home, excited, and told me to quickly run to my radio and tune in to a particular radio frequency. There was an ongoing live rescue of a sinking ship at sea, and amateur radio operators across the world were participating in triangulating the location of the distressed vessel. He wanted me to learn how to be able to do this myself in case I was ever needed. The net controller never called on me, but I did find what turned out to be the correct bearing to the ship and listened in relief as rescuers arrived.

To me, this was both science and religion in action—gaining all the knowledge I could, and then using it to help other people.

But back to Carl Sagan. As with evolution, *Cosmos* could only cover so much in a one-hour episode. Sagan discussed the Big Bang, the speed of light, and the amazing paradoxes surrounding it, how we knew the distance to galaxies barely within reach of our telescopes, how we knew the age of the Earth and the universe. Despite the limitations of sixty-minute episodes, important gaps in my knowledge were being filled in. Solid groundwork was being laid for more advanced studies that I would soon begin in college.

I'm sure I've pointed it out *ad nauseam* in this book, but it bears repeating: Sagan never told me that I *had* to believe anything I was seeing. There was no punishment if I didn't agree; no reward if I did. I was simply being given information and encouraged to go out and explore, to question, to experiment.

After years of having my sincere inquiries dismissed and snuffed out by scorn and warnings that I was behaving exactly as Satan wanted me to (and I did on several occasions hear these exact words), it was refreshing to be encouraged to simply think for myself.

Oh, so many lessons from *Cosmos*! Persecution of scientists by the church struck a familiar note. No, I was not sentenced to house arrest for the latter part of my life, as was Galileo. I wasn't burned alive, as Bruno was, for recognizing that the Earth orbited the sun. But I was strongly criticized and threatened with Hell for speaking scientific truths in church. Book bannings and burnings from the earliest days of science? Unfortunately, we still encountered that in the Bible Belt as I grew up. I made a secret pact with myself because of *Cosmos*. I would make sure in my life to read as many banned books as possible.

I found an answer to something that had long puzzled me. For all my love of the science of chemistry, beginning with the gift from my parents of my first chemistry set through all my high school studies, I never learned where the chemical elements actually came from. Sagan gave me an answer, later to be confirmed in more advanced college physics and astronomy classes: the elements were forged by the process of fusion in the cores of massive stars. We were all quite literally, as Carl said, made of star stuff. I looked at my own body with a new, almost spiritual, fascination. The iron, calcium, carbon, and other elements of which I was comprised came from the stars.

I smiled to myself in a moment of realization: I'd always wanted to travel in space. Apparently, I already had. *Cosmos*, even though just a thirteen-episode television series, would be one of the most transformational education experiences of my life. It filled in gaps in a public education that omitted important details, due to external fundamentalist pressure. It filled in those gaps with evidence.

Oh. The third significant event in my seventeenth year of life, that I mentioned at the beginning of this chapter? It was the eruption of Mt. Saint Helens. Incredibly, Young Earth Creationists in my church were already hailing it as proof that the Grand Canyon had been created by a global flood. It would be another year, when I entered college, before I had the education to understand how and why the YECs were wrong. If the reader will forgive me, I'll postpone that discussion for a few more chapters.

Regardless, after viewing *Cosmos*, the garden of doubt that had first been planted when my Sunday school teacher misinformed me about rainbows had now grown to maturity. The corn was at full height, so to speak. The contradictions between the fundamentalist biblical teachings and scientific fact were now so strong I could no longer ignore them. I felt torn by guilt. I hadn't given up on God or Jesus. Carl Sagan never asked me to. But I could no longer continue lying to myself, or allow myself to be lied to, about facts of a physical universe that I could observe myself.

It was time to make a hard decision.

Rebellion

Except for the obvious incidents common to every toddler, I'd never before stood up to my father; never told him "No." Sure, there were the inevitable arguments about bedtime, having to mow the lawn when I wanted to be out playing baseball, but I had never really stood up to him and said, "I'm not doing that."

That all changed one Sunday morning shortly before my eighteenth birthday.

The conflicts between what I was being taught at church and what I was learning to be true could no longer be ignored. The nightmares over what would happen to me for doubting what I was told about the Bible, the constant reminders from my congregation that my Catholic mother wasn't a true Christian and wouldn't make it to Heaven with me, elders telling me things I could easily prove to be wrong... it finally became too much. I couldn't take another night of waking up screaming because I saw my mother on fire.

I was so incredibly weary from the mental gymnastics of pretending rainbows didn't exist, that dinosaurs walked with humans, and the impossibility of a young Earth.

And so, on that Sunday morning when Dad came in to wake me up for church, I took a deep breath and said, "I'm not going."

I'd finally had enough.

I don't know who was more shocked, me or Dad. This wasn't how things worked in our world. Mom and Dad told us what to do, and (maybe with a little bickering) we did what we were told. End of discussion. Except on this day.

"What do you mean you're not going? Of course you're going. Get up and get dressed."

I cringed. This was the man I respected more than any other on the planet. But I held my ground.

"No, I mean it. I'm not going. I don't believe in what they're teaching there. They're saying horrible things. Things that don't make sense. Things that aren't true."

Dad wasn't going to give in so easily. "Like what? You mean you don't believe in God anymore?" His face was starting to turn red. He was mad. Mad with a capital "M."

"I didn't say I didn't believe in God anymore. I said I don't believe in what they're teaching at church."

One thing that adults never seem to learn about kids: If they've studied subjects more than the adults, then don't ask the children for opinions or explanations. Please forgive any apparent arrogance when I say it just isn't a fair fight. I'd learned this with the theology experts at church. Now, as much as I loved and respected my father, I launched into all the horrors, contradictions, and mistakes that had been building up in my mind for years.

This was an inequitable, one-sided battle. My parents had struggled their entire lives to ensure their children had the best possible educations. Recognizing my aptitude in school, I had been groomed from my earliest years to become the first person in the history of my family to go to college. My love for both science and religion had always been strongly encouraged and supported. Mom and Dad simply never had open to them the educational opportunities that would have prepared them for a debate with someone who knew the Bible so well. For an argument with someone who was also solidly grounded in biology, chemistry, and physics.

I wasn't proud of my knowledge; in fact, I was somewhat embarrassed by it. Other kids made fun of me for it. And I didn't want to argue with my father, someone I loved and respected so much.

When I finished talking, Dad just stood there in silence for quite a while. I don't believe it was in anger or disbelief over my talking back for the first time. I think he felt pain that I was rejecting his beliefs. Young earth creationism holds that if you reject any part of Biblical teachings, you're rejecting *everything*. Such was the doctrine of my church. I see it in so many of the churches of today. It's the core belief of organizations like Answers in Genesis, who insist that every single word of the Bible is true, and nothing can be disregarded.

Dad simply turned and walked away. I had won, but I had hurt someone I deeply loved and respected. I realized another truth about fundamentalism: it relies heavily on peer pressure, especially from family and friends, to keep you in the fold.

In this moment of newfound freedom, I had no feelings of victory. In fact, I had doubled my guilt. In my mind, I'd already hurt Jesus and God for daring to deny the teachings of my church; that every word of the Bible was literally true.

I'd also rejected the deeply held beliefs of my father. In hindsight, that, of my two sins, would hurt far worse, and for much longer.

Aftermath

To my relief and their credit, my parents did not press me on my decision to no longer attend church. I'm sure conversations were had; I was not privy to them. Perhaps they saw this first act of open rebellion from their son as the culmination of the religious turmoil they'd witnessed within me, going back over a decade. Maybe I would be more at peace, not so openly and obviously torn by the conflicts between religion and education. Perhaps they looked at the bigger picture: I was regarded by friends, family, and acquaintances as a good, moral person.

My mother would beam with pleasure, recounting tales of how I'd spontaneously run to help an elderly woman carry groceries to her car, or the time I got too much change back from a soda machine and was honest enough to walk back into the store and return it. I was almost a model son. I didn't drink. I didn't do drugs. I didn't violate curfews. My scores in school were well above average, and teachers praised my work ethic. Managers who oversaw my work on summer jobs would tell my parents I was one of the most honest, hardest-working kids in their crews.

In short, I really never gave my parents any reason to worry about my moral and ethical upbringing.

Unfortunately, this would not be enough for many of the evangelicals in my life. Being in a small church in a tiny town invariably meant that not only were my fellow church members my neighbors, in many cases, they were relatives. Two of my many Sunday school teachers were my cousins. It was impossible to avoid encountering such people in daily life, not to mention family gatherings.

The pressure was relentless. "We miss you at church," delivered with a smile, was always the preface to a discussion I didn't want to have (but was constantly forced to have anyway). In the early days, as I distanced myself from religion, I just didn't want to have these conversations. I learned very early that they never ended well. I wasn't out to force my beliefs on anyone. In fact, I was still exploring and learning what my beliefs actually were. But they no longer included core concepts that, make no mistake, my elders were openly telling me were going to land me in a lake of fire if I dared to not believe.

This isn't to say that I didn't also have positive Christian influences during this period of life. A beloved friend, who would go on to become a church elder, would make the decision of "coming out" and marrying their same-sex partner. This person was another who displayed Christian love, tolerance, and compassion by their deeds. In their eyes, I was never judged or condemned. Many people who had previously proclaimed how good a person I was now avoided me like the plague (except to proselytize). Yet I got the clear message from this friend that I was still loved. I was still accepted as being the same moral person I was before. I have learned many important life lessons from this Christian friend on how to treat other people.

In the future, I would later undergo a drastic transformation. Instead of avoiding conversations, I'd pause and engage with the person trying to convert me, bringing fully to bear all of the religious and scientific knowledge I had at my disposal. But, for right now, I just wanted to be left alone. I was oh so weary of the nightmares and the constant feelings of guilt.

I'm sure that in their hearts, those trying to bring me back into the fold were doing it because they cared for me and believed they were doing me a favor. I do appreciate it when people care, even if their motives are perhaps misguided.

But I was fine. I just wanted to be left alone.

Moonshine

My senior year of high school literally ended with a bang. My chemistry scores were high enough that I qualified for an elective independent course in advanced chemistry. I was the only student in the class and had one-on-one time with one of the best teachers I've ever had. Because I'm trying not to identify individuals by name unless it's unavoidable, for the sake of this book, we'll give him the boring name of Mr. Smith.

Mr. Smith had a reputation in our school. On the exterior, he was a strict disciplinarian who tolerated no nonsense in his classroom. His face seemed to be frozen in a perpetual scowl. His eyes were small and beady—they seemed to always be in motion, scanning the room. In class, I constantly felt like I was being watched to see if I was paying attention. We were in the room to learn chemistry, and for the hour Mr. Smith had us in his grasp, learn chemistry we did.

On the inside, the man turned out to be kind, with an incredible sense of humor. When he found out that members of our photography club (who used the chemical storage room as a darkroom) were making fake IDs for spring break in Florida, part of their punishment was having to make a fake ID for him as well. Mr. Smith, probably in his fifties, became Joe Jones, age 21, ready to hit the beach bars down south.

Yes, Mr. Smith did confiscate the fake IDs. He was the responsible adult in the room after all.

I was allowed to design my advanced chemistry curriculum

myself. I was handed some books on project ideas and told to pick what looked interesting. Just for the fun of it, being in Kentucky and all, I put down "moonshine still" as an option. I was just joking. We had discussed distillation the year before, and Mr. Smith even had a small bottle of peach flavored alcohol he'd made and brought out for demonstrations. No, nobody ever tried to drink it. He was a good teacher—we were well educated on the dangers of ethanol poisoning.

To my surprise, this project was approved. I should mention here that for most of this course, Mr. Smith left me completely alone in the lab. I think he appreciated the quiet time in the teacher's lounge. Here I was, eighteen years old, in complete and total possession of a full laboratory with no supervision, other than check-ins and progress updates. And I was going to build my own moonshine still.

What could possibly go wrong?

Things got off to a good start. I found a large clay pot in my family's garage and brought it to school, filled it with corn, yeast, and water, finally storing it in a dark corner of the supply room to let it ferment.

In the weeks it took for nature to do its thing on the corn mash, I played with one of my favorite lab toys—a Geiger counter. Our school was allowed to have a small amount of radioactive material on hand. It fascinated me to learn about radioactive decay... one chemical element turning into a completely different element. What I learned about half-lives of radioactive elements would later allow me to understand why my fundamentalist church was so terribly wrong about the supposed 6,000-year-old Earth. That tale will be told when my story reaches my college years.

But finally, my corn mash was ready. I took the container from its dark corner, removed the lid, and was rewarded with the foul smell of rotten corn. I struggled not to vomit. The things we do for science.

The moonshine still itself was fairly easy to construct using supplies from the lab. I poured the mash into a large round glass container that rested on an iron tripod, and placed a Bunsen burner underneath. When the mash was boiled, vapor escaped into a long glass apparatus that can best be described as a tube-inside-of-a-tube. The vapor traveled through the inside tube while cold water was fed through the outer. Ethanol would condense as the steam cooled and was collected in a beaker for later analysis by Mr. Smith.

At least, that's how it worked in theory. When the big day arrived, I lit a match, brought the mash up to a violent bubble, and watched the alcohol-laden vapor begin to creep into the system. It was time. I turned on the cold water.

I should have checked my plumbing first.

It was later described by some as an explosion, but that wouldn't be quite accurate. Teachers in adjoining rooms described it as a loud gunshot. I had misconfigured the tubing and fed the cold water right back into the boiling mash chamber. The vapor pressure and the temperature difference between the boiling mash and the cold water was just too much for the glass container.

One moment I was a moonshiner. The next, I was a forlorn high school senior standing in a room with sour corn mash dripping off the walls. Broken glass was everywhere.

I expected Smith to be furious, but he broke out laughing when he saw the destruction and realized that, miraculously, I hadn't scalded and/or cut myself to pieces. To his everlasting credit, my instructor turned the event into a "teaching moment" and had me write a paper on what went wrong. This was no bunny paper. I had to start by using the ideal gas law to calculate the approximate pressure in the container at the time of the explosion (which I still swear wasn't an actual explosion), and get more technical from there. I got an "A." Like I said, my teacher was tough but fair. And thankfully, he had a good sense of humor.

So what does all of this have to do with my journey through religion and science?

What I learned from Mr. Smith is that in science, sometimes things go wrong. You don't always get the results you expect. When you don't get those results, *you admit you were wrong*, and you revise your approach and try again. I could have believed for all the world that I was going to have a beaker full of corn liquor at the end of the day, but in the end, my beliefs didn't matter at all. I channeled Carl Sagan from *Cosmos*. Recognize your mistakes and learn from the experience.

Accepting new and better evidence, discarding that which had been found lacking, was not something I learned in a fundamentalist church. I flashed back to my Sunday school teacher's insistence that rainbows didn't exist until after Noah's Flood. Why couldn't we just admit that was just a story, and move on? It wouldn't have changed my religious beliefs at all. In fact, I think it would have made them stronger. I was taught by my church to seek truth, and I was finding truth in science.

The contrast with my faith could not have been more glaring. Young Earth Creationism and Biblical literalism would not allow followers to admit they were wrong, even if they were. When I confronted my church elders with scriptural errors and contradictions, they simply insisted God's word was always correct. If they were forbidden from acknowledging a mistake, how could they ever really tell me the truth? This troubled me deeply.

Logic, and the willingness to admit and learn from my errors, were taking on an ever-increasing importance in my life.

Part Two

Revelations

"If the general picture of an expanding universe and a Big Bang is correct, we must then confront still more difficult questions. What were conditions like at the time of the Big Bang? What happened before that? Was there a tiny universe, devoid of all matter, and then the matter suddenly created from nothing? How does that happen? In many cultures it is customary to answer that God created the universe out of nothing. But this is mere temporizing.

If we wish to courageously pursue the question, we must, of course ask next where God comes from. And if we decide this to be unanswerable, why not save a step and decide that the origin of the universe is an unanswerable question? Or, if we say that God has always existed, why not save a step and conclude that the universe has always existed?"

— Carl Sagan, *Cosmos*

College Years

My college years were relatively quiet, religiously speaking. I had stopped my regular attendance at our evangelical church shortly before entering university, though I'd still occasionally attend for special events such as seeing family members in Christmas plays. But there would be no weekly religious services for me after I first stepped on campus.

This doesn't mean that I had given up on believing in God and Jesus. The only real difference was that I was no longer surrounded by other people who were constantly re-affirming my faith, and, blessedly, I was free of the threat-of-Hell indoctrination. I was finally able to think for myself. No more church? I just didn't feel the need. But I hadn't given up on religion. In fact, some of the best religious education I ever received came from elective university-level courses.

Modern-day conservatives place a lot of blame on colleges (and secular education in general) for "brainwashing" students. Nothing could be further from the truth. Not once did a professor ever threaten me with eternal torture if I didn't believe in the subject matter. Once, in Calculus, I thought I successfully proved that one equals zero. I did get a bad grade for that one, but the threat of Hell never came up as punishment for being wrong.

As time went on, it dawned on me that any indoctrination I'd undergone in my life had occurred on Sunday mornings after the church bell rang.

I liked college so much that I did it twice. My four-year bachelor's degree was in Computer Science with a minor in Math.

I loved science so much that all of my electives were in the natural sciences (such as astronomy, geology, and biology). I also continued my foreign language studies, which would help me later in life when considering scriptural translation errors.

Less than ten years later, while doing hospital pharmaceutical programming, I found myself back at university as a pre-med student. I was enjoying working in a medical-related field so much that I fancied myself becoming a doctor. Though I learned the hard way that being a full-time salaried employee and a college student wasn't something thirty-year-olds were cut out for, two years of concentrated studies in subjects related to medicine and natural science were invaluable in helping me better understand the world.

In the following chapters, I'll be interweaving lessons learned from these two university experiences in a way that I hope illustrates how a solid secular education allowed me to make great strides in my journey to reason, significantly, without anyone once ever disparaging religion in any way. The most positive thing I took away from college is that I learned to think for myself—*I learned how to learn.*

Astronomy 101

Unlike most college freshmen, I knew exactly what I was going to study from the day I enrolled, and I never deviated from my chosen major. It was always going to be Computer Science. Anything else was out of the question. Despite the name, this field wasn't so much science as engineering. Though we applied scientific thinking and logic, this wasn't considered a *natural* science, like biology or geology. But because I loved natural sciences so much, I filled my electives with as many of these courses as I could.

Astronomy was an absolute no-brainer of a choice. I'd been staring at the heavens in wonder since I was a small child, and my first telescope, that wonderful, treasured instrument, was still safely stored at my parents' home when I moved into my dorm room. I would now have larger, more powerful telescopes at my disposal thanks to my university.

Astronomy 101 was a two-part course. We had lectures during the day, and, of course, mandatory time at the university's observatory at night. A fourteen-inch reflector was the showpiece of our small observatory, but for nightly observing assignments, smaller Celestron four-inch scopes, shared with a lab partner from class, were more than enough. Even these scopes had incredibly more light gathering power than my small (but beloved) starter scope.

Yet it wasn't under the still (mostly) light-pollution-free Kentucky skies that I'd experience my most meaningful lessons in astronomy—that would occur in the classroom.

Lectures began with a brief history of the universe and, of course, the inevitable question, "What caused the Big Bang?" was asked on the first day of class. But this was the first time I'd heard the question asked in a non-theological setting. It wasn't a setup for the inevitable "well it had to be God" answer. This was a student asking the question in pursuit of knowledge. The answer was refreshing, and made me trust in science even more:

"We don't really know," the professor said. "Science is still working on that. But we do have undeniable evidence it happened." He went on to explain the Cosmic Microwave Background. As the famous physicist Brian Cox would later say in mock frustration, with a TV producer bleeping his hilarious challenge to Young Earth Creationists: "Look! You can f----ing see it!"

I was just in the second decade of my life at this point, but this way of thinking had great appeal to me. I'd already had so many experiences with people in a church setting giving me answers that I later found out didn't stand up to the scrutiny of a child. If they'd just told me in Sunday school, "we don't really know, but we're still searching for an answer," life probably would have gone much differently for me. I may have stayed in church. There's no faster way to lose trust than giving a wrong answer and then *insisting* on its correctness.

The most memorable moment of Astronomy 101 was when we retreated into a darkened lab, with only faint red light to guide us, protecting our "night" vision. Each workstation was equipped with the object from my early childhood that would capture my imagination—and initiate my skepticism—forever: a prism.

The prism from my first-grade science class. The prism that broke white light into its components. The prism that originated the argument between me and a Sunday school teacher over the non-existence of rainbows until after the great flood. The prism I'd recreate in countless water drops in my backyard with a water hose, trying in vain to NOT get a rainbow, as my church told me had once been the norm.

We did something special with our prisms in astronomy class. We didn't just split white light into a rainbow spectrum. We split light from special bulbs that contained gases such as sodium. We still saw partial rainbows, but something startling appeared before our eyes: there were dark lines in the spectrum (the color spread of the rainbow, the refracted light).

Our professor explained that these dark lines were caused by the chemical elements in the bulbs absorbing light at a given frequency (or wavelength). Imagine an old FM-style radio tuner in your car, picking up stations from 88 to 104 megahertz. This was part of a *spectrum*. Now imagine that suddenly, let's say at 90 megahertz, where your favorite Rock station would normally appear, you got nothing but silence. What if something was there, absorbing the emissions at this frequency?

This was exactly what we were seeing with our own eyes. Visible light, and radio signals, are all part of the same electromagnetic spectrum. It just so happens that we can see a small part of this spectrum with our eyes. Other parts are invisible, and can be revealed by radio receivers or, in the case when we thought I had a broken arm, an X-ray machine.

The important point here was that I was able to simply look at a light source through a prism, and I could tell you what that light source was made of. It didn't matter if the source was a light bulb there in our darkened laboratory, or (here's the fun part) light traveling billions of years through space before reaching Earth. The dark absorption lines always appeared at the same place in the spectrum, like fingerprints.

Except that they really didn't. There was a catch.

Most of us are familiar with a phenomenon called Doppler Shift, even if we don't know it by that name. The sound of a fire truck's alarm seems to change in pitch as the engine approaches us and then passes by. The same for a speeding train sounding its horn at a crossing.

And so it would be for light from distant stars and galaxies. Yes, the dark absorption lines from hydrogen, the most common element in the universe, were still there when we looked at light from distant objects, but the lines were shifted to the right or left in the spectrum, depending on whether the star/galaxy was speeding toward us or away from us.

A brilliant astronomer named Edwin Hubble showed that the shift of these dark lines I was seeing with my own eyes was proportional to the distance to the stars and galaxies emitting the light, and how fast they were moving toward or away from me. When I looked up at the night sky, I was seeing light that had left those stars and galaxies hundreds of thousands, or even billions, of years before.

I'd always been suspicious of, to the point of completely disregarding, my early church's claims that the Earth and the universe were 6,000 years old. Now I knew I was being outright lied to. I'd like to be a little less confrontational in stating this, but I think it's important to speak honestly: the moral authorities I'd entrusted my life and soul to as a youngster were wrong. A 6,000-year-old universe was not possible.

I also could now see with my own eyes that my Sunday school teacher did not tell me the truth all those years ago when she said there were no rainbows before Noah's Flood. We were splitting light from stars that left thousands, hundreds of thousands, millions, even billions of years before this flood allegedly occurred—and yet the familiar rainbow was still there.

Fundamentalism does great damage to its integrity when it insists on absolute truths, and those truths don't stand up to scrutiny. When someone fails to tell you the truth enough times, you eventually stop trusting them. My religious mentors went to great lengths to convince me that everything I would learn in the secular world was the result of evil influences tugging at my soul. Yet there I sat in a lab, staring at incontrovertible evidence they were wrong. To this day, it is hard to understand why they were (and still are) so insistent on denying basic science. Why was the age of the Earth so important to them? Many religions had absolutely no problems with these scientific facts.

At no point during any of my astronomy lectures did our professor ever bring up theology. He never told me I had to believe something. Unlike church, I was never threatened with damnation or the loss of paradise if I didn't agree. The facts were just laid out bare for the class. There was no indoctrination. I was free to believe or disbelieve what was right there in front of me. I could have turned off the light bulb and made the spectrum and absorption lines disappear.

But I left that light bulb on. It was a powerful metaphor for having gained knowledge.

Geology 101

Introductory astronomy had completely shot down any lingering thoughts of a young universe. It did so without ever mentioning religion, but did cause a good deal of mistrust in my theological upbringing. This occurred simply because I had grown up being taught that truth was an essential part of being a good person, and I was now able to look with my own eyes and see that Young Earth Creationists at church weren't giving me the whole story. This realization did not in any way go against my closely held belief in a creator. No professor ever spoke against God.

It was interesting that a simple prism had raised the first doubts in what my congregation was teaching me, even at an early grade-school age, and I'd come full circle to a prism in a college astronomy class confirming my doubts in certain dogma. College itself wasn't casting shade on what I'd learned in church—the church was doing that itself, by telling me things I'd repeatedly find out just weren't true.

My first college-level geology course would further drive this point home. Geology was another natural science elective I wedged into my schedule, just because I was curious about the Earth. Our professor was a kind, older gentleman, clearly nearing retirement age, who taught in the monotone voice from a famous 1980s movie: "Bueller. Bueller. Bueller."

Many lectures involved the professor loading up an old-style film projector, flipping the "on" switch, and then letting the class watch something akin to a National Geographic special while he would nod off at his desk. Some of the students took advantage of his *apparent* inattention to goof off. I say apparent because our exams were thorough rake-you-over-the-coals grillings on the content of the films. This teacher was clever. He could apparently see with his eyes closed.

The first significant topic I recall concerned the effects of erosion on the Earth. We compared and contrasted slow, meandering rivers to fast-rushing rapids and, very meaningful to me, catastrophic flooding. I'd always been taught in church that the Grand Canyon was formed in one huge flood. I questioned this, much to the anger of my Sunday school teachers, but with no real secular education in the matter, I couldn't really debunk my elders. One hour in geology class would change everything.

As the film projector clacked away incessantly, we saw the effects of water meandering slowly over a landscape: long, random curves, and "oxbows," where the entire river would literally double back on itself. We saw imagery of this effect in eastern rivers, and in western rivers, including the Colorado River, which wound its way through the bottom of the Grand Canyon.

This three hundred mile long canyon, supposedly carved almost instantaneously in one massive flood, was full of zigzagging curves and oxbows. In one short section, the Colorado did indeed make a beautiful, complete loop as it ceased its (roughly) westward journey, headed ninety degrees off to the north, made a graceful curve that eventually had it heading (approximately) ninety degrees to the south, and then yet another bend that sent it roughly on its original westward course.

Next, we looked at both fast-moving rivers and straight-line flood damage. No gentle curves. No oxbows. We could have been watching this film over a decade earlier in grade school and the entire class could have seen the difference. The Grand Canyon couldn't have been carved by a rapid flood. As we continued geology studies, this one piece of evidence from erosion would be backed up by many more facts, but this fact was already enough.

Once again, I hadn't been told the truth while growing up. It's important to point out that in 2024, with so many fundamentalist/conservative pundits claiming that college is a place for indoctrination and leading people away from God, our instructor never once woke from his apparent nap to disclaim a biblical flood. We just sat there and watched a film on the power of water and the effects its force had on the planet.

This would be a recurring theme as I earned my university degree, and I'll mention it often in this book because it's so important: I was simply given information and allowed to think for myself. Indoctrination? No. Not at my university.

We would touch on a topic that would lay the foundation for a better understanding of a hot topic that would come up decades in my future: climate change. I learned, to my surprise, that most of the carbon dioxide in the planet's atmosphere came from natural sources, such as volcanoes and wildfires. I was shocked. Even this early in my life, climate scientists were already sounding alarm bells about how the Earth was changing.

But then we got to the part that isn't taught on Facebook or YouTube videos produced by oil companies. Over long periods of geological time, the planet had achieved a balance with carbon dioxide. The amount of the gas that was naturally produced was offset by natural "sinks" in the planet—our oceans and our forests. Over timespans unimaginable to humans, we had reached an equilibrium. Humans were changing that.

We were now, since the beginning of the Industrial Revolution (through the burning of fossil fuels), emitting more carbon dioxide than the planet "knew" how to handle. We were demanding more of the Earth than it could do. This was a fact I'd tuck in my back pocket for future use, when climate change denial became so prevalent in the 2020s. Now, as back in my university years, fundamentalists were on the front lines, claiming that God wouldn't possibly create a planet that humans could adversely affect. They'd bring up the fact that volcanoes were producing more CO_2 than humans.

I had a factual answer ready, but nobody would listen.

Another memorable moment in Geology 101 was when we talked about how old the Earth was, and how we came to that conclusion. In a section on volcanic activity, radiometric dating was explained. I already knew from Mr. Smith's chemistry classes in high school that some elements were radioactive, and decayed over time, becoming other elements. Aside from building moonshine stills, playing with a Geiger counter was one of my favorite pastimes when left alone in the lab.

With apologies to any geology/physics/chemistry experts who may be reading this book, I'll try to explain in a manner that's not too "geeky." Once again, via film, our class watched a volcano erupt in Hawaii. Lava flowed down a mountainside. The narrator explained that the heat of the lava drove off volatile gases, such as argon. Argon would turn out to be key here. In addition to being volatile, it is also remarkably chemically inert. It doesn't readily combine with other elements. As the lava cooled and solidified, argon would be driven off. Other elements—solids, those that reacted with others to form compounds—would remain. One of those often turned out to be radioactive potassium.

Something surprising happened when these rocks were later broken open in the absence of air. I mention air because, in addition to being comprised mainly of nitrogen, with a fair amount of oxygen, the air we breathe also contains trace amounts of argon. Remember, I said argon would be important.

So what was found inside these rocks? Well, of course, there was radioactive potassium. But amazingly, there was argon. Not just any type of argon. "Radiogenic" argon—produced by the decay of another radioactive element. How did it get there? It shouldn't be present. The volcanic heat would have driven it away.

It turns out, the film's narrator explained, that argon is a decay product of radioactive potassium. In an earlier chapter, I mentioned the term "half-life" as I described playing with a Geiger counter in chemistry class. This term refers to how long it takes for one-half of a given sample, say, radioactive potassium, to decay into another element, like argon. The half-life of potassium-40 is approximately one and a quarter *billion* years. Given a sample of rock in which we knew we only had potassium to begin with, and knowing that any argon we found had to have come strictly from its parent potassium, we could look at the potassium-argon ratios and figure out how old the rock was.

The planet wasn't 6,000 years old as we were taught in church. It was billions of years old. 4.6 billion years, to be more precise.

Scientists didn't rely on just one rock found in one lonely spot on Earth. This result was reproducible across the world. And, our very interesting film went on (over the light snoring sounds of the instructor), science didn't just rely on potassium and argon. There were many other radioactive elements, all with different half-lives, many of which overlapped. Ignoring all the other lines of evidence I'd eventually come across to affirm we lived on an ancient planet in an ancient universe, volcanic rocks alone were enough to remind me that, once again, I had not been told the truth by religious elders.

We went on to learn much more about radiometric dating; technical details that I fear would bore the reader. To summarize, scientists didn't just walk out in a field and chuck a rock into a machine to get a date. Geological context was important. This process didn't work like a pregnancy test, with a near-instant result. And there were careful checks against contamination. Interested readers might want to check out "isochron dating."

In yet another class, we'd finally come to a study of the geology of the Grand Canyon itself. It would be a few decades before I'd visit this natural wonder in person, but even in photos and film, it was obvious that the rock walls of the canyon had been laid down in homogeneous layers—one type of rock clearly separated from another type of rock. Having already studied the basics of rock formation in earlier classes, I knew that a lot of this stone would have taken large amounts of time, on geological timescales, to form.

I recalled a church sermon from a traveling Young Earth Creationist preacher who claimed that all of these layers were laid down at once in a massive deluge. That made no sense anymore. We'd studied what happens to debris and material carried in floods. To be very unscientific, it all ends up as a jumbled mess. Other than a process called hydrological sorting, where larger, more massive chunks of material tend to sink to the bottom faster, my meager Geology 101 education showed me that the traveling preacher was wrong. There should have been a planet-wide debris field, a heterogeneous mix of everything (including all life) that was swept away in Noah's flood.

No such debris field has ever been found.

I mentioned in my chapter on *Cosmos* that the eruption of Mt. Saint Helens was one of the significant events in my seventeenth year of life. Young earth creationists had made this volcano an icon of their beliefs, pointing to a "mini Grand Canyon" carved into the side of the mountain by a flow of mud and volcanic debris. In college geology, I learned the myriad problems with these claims.

First, the "canyon" on Mt. Saint Helens was perhaps a hundred or so feet deep, depending on where you measured. The Grand Canyon averages a mile in depth.

Second, the Grand Canyon was carved from solid rock. The volcano? No. The Mt. Saint Helens mountainside was a collection of loose rock, dirt, and volcanic debris.

Third, as I mentioned earlier, our geology professor pointed out that catastrophic events usually produced straight-line damage, as opposed to the slowly-carved, meandering oxbow curves of the Grand Canyon. Aerial and satellite footage of the "canyon" on Mt. Saint Helens clearly showed a huge gouge with barely any curves—certainly nothing found in the Grand Canyon, where the Colorado River literally doubled back on itself.

Young earth creationists would not give up on the volcano, though. It seemed to take on an almost spiritual meaning to them. One YEC "geologist," Steve Austin (not to be confused with the fictional character from the TV show *The Six Million Dollar Man*), would use the mountain to attempt to disprove the radiometric dating methods discussed earlier. He violated every rule mentioned pertaining to obtaining samples in proper context, and grabbed some one-year-old rocks from the volcano. He knew the age of the rocks because the lava had only recently solidified. He then sent these samples to a lab that warned him beforehand that the equipment they used was incapable of dating anything younger than approximately 200,000 years. Austin ignored the warning, came up with a false result, and proclaimed to the world that radiometric dating just didn't work.

Had Austin simply sat with me in that class, he would have known, as our professor told us, you don't just pick up a random rock and stick it in a machine to get a result. In fact, as Austin apparently did have a degree in geology, he would have already known this. So why the subterfuge? It was becoming apparent to me that Young Earth Creationists really had no method of scientifically dating anything at all. Their *raison d'être* was to simply deny and attempt to disprove everything that real science was doing.

This type of dishonesty would become a hallmark of those who believed in a 6,000-year-old planet. YEC paleontologist Hugh Miller (not to be confused with the famous Scottish scientist of the same name), misrepresented himself to a museum in order to obtain fossilized dinosaur bones which he intended to date using the carbon 14 method. As we would learn in school, there were a lot of problems with this. First, fossils would mean that the original bone had been replaced by minerals. So you wouldn't actually be dating bones. Further, both our chemistry and geology teachers had made it plain to us that most minerals don't contain carbon. Some do, but you wouldn't expect carbon 14, a radioactive isotope found in Earth's atmosphere and taken in by living things, such as plants, later to be ingested by animals.

But, worst of all for Miller, museums used to cover fossils in shellac, an organic substance produced by beetles. The misguided goal of early museums was the preservation of the fossils.

And so, when Miller presented his "dinosaur bones" to a lab for dating, he ended up getting back a date for the only organic substance present: the shellac in the bones. Not surprisingly, the date was not congruent with the known age of dinosaurs. Like Steve Austin, Miller triumphantly proclaimed that he had either disproven another dating method, and/or dinosaurs had lived much more recently than science claimed.

Throughout my university education, and ever since completing it, I have looked in vain for a single Young Earth Creationist method for scientifically dating anything. They will sometimes concede to tree rings, perhaps because they know it's hard to find trees older than their 6,000-year-old claim, but science has been clever enough to match overlapping growth patterns in trees that lived in the same environment at the same time (some older, some younger). By aligning the ring growth patterns in these trees, science has gone well beyond the YEC 6,000-year age for the planet.

All of these very simple facts, I learned via a basic secular science curriculum, where not once was the subject of religion ever mentioned. Why were followers of fundamentalist religion so obsessed with the denial of basic science?

The answer, of course, was that to deny a single word of scripture was to deny all of scripture. And that just couldn't be allowed.

Sadly, even as the manuscript for this book leaves my hands and goes to my editor, my adopted home state of Kentucky has proposed a constitutional amendment that would require taxpayer dollars to fund religious schools, with little oversight over what children will learn. I have serious doubts that if this comes to pass, any science or history that contradicts my state's strict fundamentalist government will ever make its way into a school. In my youth, overwhelming pressure against basic, essential topics such as evolution kept the subject out of our classrooms.

I worry that, at this rate, there will soon be no science or history left.

Brother Jed and Sister Cindy

College was a breath of fresh air, religiously speaking. I no longer had the constant peer pressure to believe. That doesn't mean I had given up yet on God or Jesus. I was just left alone to think for myself. In hindsight, I can see that religions survive, at least in part, by the constant reinforcement and peer pressure of being in a like-minded group. That was certainly the case for me and my dozen-or-so years of weekly church attendance.

Contrary to current-day right-wing media, nobody in college ever approached me and tried to change my mind about religion. Except for the theology courses I intentionally took, theology was a subject completely absent from the classroom.

That doesn't mean the topic didn't come up at all. Our campus was a favorite hangout for missionaries and evangelical preachers. The pair who were so well known that they now have their own Wikipedia pages were Brother Jed and Sister Cindy. If you went to college in the South or Midwest in the 1980s, they need no introduction. But for the rest of you:

Brother Jed and his wife Cindy were pure fire and brimstone, and they believed in direct confrontation as the only way to bring people to Jesus. They'd set up in our free speech area, commonly known as "The Quad," every month or so, and begin harassing students using bullhorns or battery-powered public address systems.

Cindy would latch onto girls wearing sorority shirts and follow them down the sidewalk, screaming that they were painted whores whose only purpose was to lay on their backs and spread their legs for fraternity brothers. Brother Jed would back her up by shouting what he thought were applicable Bible verses into his microphone.

Obviously, when you provoke a crowd of free-thinking college students, you're going to get the confrontation you asked for. The sad thing is, many of these students could have already been Christians (or Jews or Muslims) themselves. But nobody wants to stand around and be insulted.

I was long past the days where I felt I needed to preach to everyone, a complete reversal from my early Christian days. I was still at the point where I would share my beliefs with someone who seemed curious.

But after Brother Jed and Sister Cindy?

I wouldn't have spoken to someone about Jesus if you held a gun to my head. These people made me embarrassed to be a Christian. Extremely embarrassed. I could hear the derogatory shouts from those in the crowd who were non-believers. Even if I had wanted to, I had no way of approaching these people. Not after the intro from the two screaming evangelicals.

Inevitably, students would set up stereo speakers in dorm rooms overlooking the crowd and start blasting Van Halen's "Running with the Devil" or AC/DC's "Highway to Hell" at full volume. Jed and Cindy tried to compete, but they couldn't shout over the rock and roll.

When there was finally nothing left to see or hear, I'd shrug my shoulders and walk away, humming as I went. I may have been a Christian, but Van Halen and AC/DC were two of my favorite bands. Despite all the warnings of evangelicals at the time, I never once felt the urge to make a sacrifice to Satan while listening to those songs. Despite the harsh warnings of my church that Rock music was leading the world to Hell, my parents gave my sisters and me wide latitude when it came to music. I was a Pink Floyd fan at age ten, when Dark Side of the Moon debuted. First taken in by the prism (prisms!) on that wondrous album cover, I became immersed in a style of music that would influence me for life.

Satanic rock and roll? I think even my parents laughed at the thought that music could have supernatural influences. They'd lived during the birth of Rock and had heard the same empty warnings from their own church elders. John Lennon's "We're more popular than Jesus" comment, which enraged churches at the time, meant little to Mom and Dad. They still loved the Beatles. I never feared my parents going through my music collection and finding AC/DC, Led Zeppelin, Black Sabbath, Pat Benatar, or other "forbidden" groups.

They just don't make music like that anymore.

The takeaway is, I don't think many evangelicals realize how much harm they're doing to their own religion. If I wanted to impress an important point on someone, the last approach I'd consider is yelling insults at them through a bullhorn.

Spring Break

College spring break is a rite of passage for many American students. For those of us on the east coast, this usually means a ubiquitous trip to the warm sunshine of Florida. Kentucky is a wonderful place to live nine months of the year, but winters are unbearable. As our climate slowly, noticeably began changing, we didn't even get snow anymore—just three months of forty-degree temperatures, constant clouds, and cold rain.

Although this part of my journey may seem to start out as nothing more than a wild tale of a young man "cutting loose" for the first time in his life, there is, for me, a significant moral and ethical lesson at the end. I was to learn something about myself, and my fundamentalist upbringings. I ask the reader for his/her forbearance as I launch into the story.

And so it was that in the year 1982, when a group of buddies suggested carpooling to Fort Lauderdale in early March, I was onboard immediately. My parents worried a little (they'd seen television coverage of the nightly revelry on Miami-area beaches, but to their credit, they trusted me). I was given the go-ahead. The trip was on.

I should point out that at this point in my life, nearing the ripe old age of twenty, I had never tasted alcohol, other than the surreptitious sips of "apple beer" Uncle Bill would sneak to me as a kid and, believe it or not, once-weekly sips of a tablespoon of homemade wine from my parents. The belief at the time was that red wine enhanced the iron content in our blood, so, every Monday night at bedtime, my sisters and I would line up and take a medicinal dose.

I was also, I'm embarrassed to say in this day and age, still a virgin. My church had been very clear on the evils of both alcohol and sex. I'd participated in neither, due to the weekly nightmares of eternity in Hell always in the back of my mind. Any time I had "impure" thoughts, even as an adult, I felt all of the guilt that my congregation had mentally beaten into me.

Some things were going to change on this trip. In addition to contributing to our daily sunburns on the beach, my friends wanted to experience something new, something unique, in Florida. I had just the thing: the game of Jai Alai. This was a sport popular in Latin America. Think of handball played with a goat-leather wrapped wooden ball, players trying to capture the ball as it ricocheted at incredible speeds off the three walls of the *fronton*. (The place for the fourth wall was replaced by a large net, so the betting crowd could see the action.)

I knew Jai Alai because it was an old family tradition to take eighteen-year-olds out gambling on their special birthday. My parents had introduced me to the sport during a high school spring break years earlier. It may seem a contradiction that a devout Christian would gamble, and that his parents would actually play a part in it. But recall that in my Catholic upbringing, I played bingo in the church basement after Mass. My first religion had no problem with gambling. And, being a Kentuckian, I grew up in the horse racing capital of the world. This was part of my culture. It was not uncommon to see church members at the race track. Even fundamentalists will bend the rules now and then. My friends were agreeable to a Jai Alai adventure, and, as the night went on, they became more excited because, like the good Kentucky gambler I was brought up to be, I could pick winners.

My suggestion was a smash hit. The winnings piled up. At the end of the night, those of us who had entered with the princely sums of $50 or so, intended to cover expenses for our entire trip (remember, these are 1982 dollars), were leaving with six or seven times that amount. Among the Florida spring break revelers, our little group was suddenly as rich as Croesus.

It was inevitable that my friends, many being college fraternity brothers, insisted that we go out for drinks to celebrate. Nobody checked IDs in Florida in those days. Our hotel bar certainly didn't. I had no experience drinking; my church told me it was an absolute sin, but lessons drilled into me early and often by my realistic parents led to an unbreakable promise between us. They knew that someday I would drink. I swore I would never get in a car with anyone who was drinking. If I ever did drink, it would be in a safe environment. I would keep that promise.

This isn't to say I didn't feel guilty as I walked into the hotel bar. I could hear in my imagination a chorus of Sunday school teachers singing, mournfully, "You'll be sooooory."

But I entered anyway... after honoring the promise to my parents and insisting that our group's car keys were locked in the hotel safe.

Incredibly, when others in the bar discovered our good fortune, we weren't allowed to buy drinks for ourselves. I don't know if we were good luck (I did give out a lot of Jai Alai tips to folks who promised they would be attending the next night), or if there were just kind souls out there who appreciated the excitement and wide-eyed wonder of college kids experiencing the world. But the alcohol flowed freely. My offers to buy for the others were flatly rejected. All that was asked was that some people (seriously) wanted to rub my shoulder for good luck. Maybe tomorrow night they'd be the big winners.

I started with something I think was a Tom Collins or a Vodka Collins. (I'm pretty good at the game "Jeopardy," but when the Potent Potables category comes up, I keep my mouth shut. I know nothing about liquor.) I didn't like the taste of alcohol; I think the bartender compensated not just with extra sweet & sour, but also the innovation of powdered sugar sprinkled on top. Based on the amount I would drink that evening, he was also clearly doing me a favor with very short pours of alcohol.

But I'd found an alcoholic beverage that I liked. I consumed maybe four. Perhaps eight. Who was counting? That sounds like a lot, but, again, in retrospect, the bartender clearly knew he had a newbie on his hands and was limiting my intake.

Inevitably though, the night slowly turned into a slow-motion blur. Most of my friends had already had enough and headed back to our hotel room to sleep. I stayed behind. I was really enjoying myself. Socially awkward, I was the poster child for the computer "geeks" often seen in 1980s movies. But with the courage alcohol gives, I became, at least temporarily, an outgoing person who people seemed to enjoy talking to. And they kept buying me drinks. The funeral dirge in the back of my head, sung by my former church teachers, was replaced by the sounds of Van Halen blaring from the jukebox. (A jukebox? Yes, as I've said, I am that old.)

Finally, the bartender made the announcement, "Last call!" I had to ask my newfound friends what this meant. I had really never been in a bar before. Oh, drat. No more drinks.

I encountered another new experience that knowledgeable readers will be familiar with: I didn't feel a thing from all my drinking, other than slight happiness... until I made the unfortunate decision to quickly stand up.

Whoa! I was no longer an alcohol virgin.

Everything was spinning. Walking a straight line was impossible. I looked down at my feet and asked, out loud, "What in the world is wrong with you?"

The most stable object within my field of vision were the loungers by the bar side pool. I collapsed into one. It was actually a beautiful night, with temperatures in the 80s, palm trees swaying in the breeze. I was actually enjoying myself. I looked up at the few stars visible in the light-polluted South Florida skies and began to think about astronomy.

Then an absolutely gorgeous girl walked up from somewhere behind me and collapsed, as I had, in an adjoining lounger. She smiled, and we began to talk. Half our words were probably inebriated gibberish, but it didn't matter. We seemed to understand each other, and were having quite the intimate conversation.

And then the clouds moved in. Raindrops began to fall. I looked over at my beautiful drinking buddy. She had passed out. I gently shook her awake.

"It's starting to rain. We need to get up."

She looked into my eyes. "Would you like to come back to my room?"

Did this mean what I thought it meant? I had a decision to make.

This girl was clearly drunk beyond the ability to consent. She could barely walk. I had three younger sisters. I could see them in this girl's face. What if it had been my siblings, practically helpless, with a total stranger standing there? I would have hoped that whatever guy they were with would have appreciated the situation and had some moral introspection. This just felt *wrong*.

I walked her to her room. She gave me a longing look as she stood in the doorway. I felt my resolve begin to waver.

No. I wished her good night, and made sure she was locked in for the evening. I began walking the streets of Fort Lauderdale looking for a McDonald's. It was 3:00am. I had the munchies.

My story may sound trivial, even ridiculous to some. Kids went on Spring Break to get drunk and have sex. Wasn't that the whole purpose?

But this is one of the moments I point back to in my life when I'm told emphatically by theists that non-believers such as myself have absolutely no morals; we don't get our direction from God, so we feel we're free to do anything we want. In debates with fundamentalists, I've been accused of lacking any subjective reason to not have done the wrong thing. I've even had these same debate opponents claim that without God, they would have gone into the girl's room. They argue that, without their religion, why not take advantage of the young woman? What's to stop them?

Sorry, but taking advantage of an inebriated college student wasn't in my moral wheelhouse. I made this decision on my own, no supernatural guidance required. I humbly submit that if your only reason for not raping someone is that you're afraid God will get mad at you, then perhaps you should examine your own beliefs. If your only reason for not taking advantage of a helpless individual is the expectation of a heavenly reward, maybe it's time to re-evaluate your thinking.

I'm sad to say that had things gone differently, my own church would have blamed everything on the woman. This is not hyperbole. I've actually sat through sermons where it was the female who was blamed for getting raped. Remember, Lot was not chastised for getting drunk and having sex with his daughters. The females were the villains in the story.

On this night, I made three decisions, all of my own volition, with no supernatural guidance. I decided to drink. I decided to gamble. I decided to guide a young woman safely back to her room.

It had been a great night. I felt good about all of my decisions. On a breezy night by a Florida poolside, I realized that my morals didn't come from an ancient book. They didn't come from my evangelical church. They came from the teachings of those important in my life, such as my parents. My morals came from the world around me. They were driven by the desire to treat others as I would like to be treated. As I would have others treat my loved ones.

I was growing up. I was learning to make decisions that would allow me to be proud of and rely on myself, rather than worrying about pleasing entities I couldn't see. In her book, "Leaving the Fold," Dr. Marlene Winell would touch upon this subject when relating stories of patients she'd helped through the deconversion process. Fundamentalist religion tends to teach that we're not worthy; that we can only be saved via a supernatural entity. We look outward instead of inward for confirmation that we are doing the right thing. Many patients who left religion would find themselves feeling as if they no longer had a moral compass.

I certainly felt this way throughout much of my Christian life. Simple, seemingly inconsequential incidents such as my Spring Break encounter were teaching me to trust in my own decisions. There would come a time in the future when I would have to seriously question whether or not I could really leave religion behind. Incidents such as the one I describe in this chapter would make it easier in the future. I'd be better equipped to cope with the trauma of disavowing most of what I'd been taught had made me a worthwhile person in the first place.

Religion 101

University was the first chance I had in life to do religious studies in a scholarly manner, without anyone standing behind a lectern or at an altar, preaching to me. I was now experiencing *teaching*, not *preaching*. Hoping to expand my knowledge, I enrolled in a series of classes on world religions.

My first course was Judaism. Interest in this religion was mainly due to a very good friend I made early on at university. Because of my sheltered life in a small town where even the one non-Protestant church was frowned upon as "not really being Christian," I didn't have any exposure to any of the other major religions.

As I mentioned in an earlier chapter, my early education on Judaism consisted mainly of a yearly Easter sermon where the Jews were inexplicably blamed for the death of Jesus, followed by a radical 180-degree turnaround thanks to televangelists like Pat Robertson and Jerry Falwell, who assured us all that Judaism and the state of Israel were essential to Christianity. Yes, those were confusing times.

Having a good friend that I could talk to openly about a different belief system was an eye-opener, and I highly recommend this to everyone. One of the wonderful things about Judaism is that it doesn't proselytize. Nobody's claiming it's the one true religion, and they're happy to let everyone believe what they want. This, in and of itself, was a breath of fresh air. Having abandoned my own ingrained need to push Christianity on everyone, I never preached or tried to convert my friend. We talked about our differences, but it was in a spirit of learning.

Our Judaism instructor wasn't a rabbi; rather, he was a professor at the university who had a degree in religious studies and just happened to be Jewish. I came to appreciate that his wasn't just a religion, it was a culture as well. Having become weary of religious rituals, I didn't identify strongly with those of Judaism. But I felt a very strong affinity to my perception of the core values of this religion. There were no threats of Hell or promises of Heaven. Do good for good's sake. Importance was placed on family and education (strong values in my own family). And, incredibly, it was okay to question God. Judaism didn't prohibit asking questions. It encouraged doing so.

When we came to the Holocaust section of our studies, I was deeply moved. Like evolution and slavery, the Holocaust received barely more than one or two lines of text in my school textbooks. I had some familiarity from watching documentaries on PBS as a kid, but nothing could prepare me for the hard truth. We traced the beginnings, from German antisemitism, all the way to actual films from concentration camps.

The antisemitism reminded me all too strongly of the Easter Sunday sermons I endured as a youngster. Later, when I would independently come across a 1939 German census (WW II began in 1939), it did not surprise me to learn that over 95% of the population were Christians. Perhaps my church wasn't the only one victimizing Jews.

We moved on to the forced internment in ghettos, closing of shops and confiscation of property, stripping of civil rights, the yellow Star of David on clothing, and then, finally, deportation to the camps.

There were tears in our classroom during these sessions. I managed to hold my own in until I watched a film of Jewish prisoners being forced to line up in front of a large pit, feet on the edge, and then machine-gunned so that the bodies would fall in without anyone having to take the trouble to move them. While this happened, other prisoners were forced to stand in line, awaiting their turn to be shot. I finally broke down. How could anyone do this?

The answer, of course, was to dehumanize them. I see this worrying trend of dehumanization of others even today as I write this book. I can trace the practice, at least in part, to words I heard on Easter Sunday, or during anti-LGBTQ sermons in my own church.

After the execution video came a shot of a stack of naked, emaciated bodies, the word "Polak" (a German slur for Polish people) written across the corpses in black paint.

I was Polish.

Many decades later, I traveled across Germany. It's customary there, when entering a train cabin, to greet everyone. An elderly gentleman noticed my accent and asked if I was from America. "Yes, I am," I replied.

He stood up, took off his cap, gave a small bow, and thanked me and my country for trying our best not to hit German churches during our bombing campaigns during the war. His family had fled there knowing they'd be safe. Thinking back to my Holocaust studies, I wanted to ask him how many synagogues on his block had been intentionally burnt, but I wasn't that cruel. I know that a large number of Germans wanted no part of that war. But I also wondered how many German Christians had heard antisemitic sermons such as I had? How many stood quietly by and said nothing when the bigotry and hatred began to spill over into their government?

I learned that this religion also had divisions similar to what I'd seen in Christianity and Islam. There were not just Jews, but Orthodox Jews, Conservative Jews, and Reform Jews. It's always struck me as odd that across seemingly all world religions, people who proclaim the same faith can have so many disagreements.

Judaism studies finally concluded and we moved on to Catholicism, the class taught by a very affable young priest. Having started out as a Catholic and maintaining close contact with my family members of that faith throughout my life, I can't say that I picked up a lot of new information here. But one new revelation shocked me.

Having never been confirmed, I wasn't allowed to take communion when I attended Mass with my family in Wisconsin. So I somehow went through my Catholic life without ever hearing about *transubstantiation*—the belief that during communion, the wine actually turned into the blood of Jesus, and the wafers literally turned into his flesh. I was appalled. No matter how hard I tried to escape it, in Christianity, the need for the spilling (and now drinking) of blood was always there.

To be fair, in early Jewish worship, God demanded animal sacrifices, but those had gone away when the Temple was destroyed. I wasn't about to give a free pass to Judaism either. My rational mind couldn't reconcile a loving, all-powerful creator, requiring the spilling of blood.

A lot of our class's questions for the priest were those I'd asked as a kid and already had answers to: why were Christians praying to saints instead of directly to God (why not skip the middleman)? What was up with all the idols (statues)? Was there any real difference between holy water and regular water? Where in the Bible did it mention purgatory? How could babies be born with sin already on their heads when they didn't even know what sin was?

One question our priest stumbled over would be asked again in my next class (Islam). This concerned the equality of women in the church. I'll correlate his response when I discuss the Muslim answer.

It was a good refresher course, and, much to their credit, the class asked questions and listened to the answers with respect. I enjoyed this kind of theological education. Never did a lecture end with an invitation to walk to the front of the class and be accepted into the one true religion.

Islam would be the third and final piece of our curriculum. This was the most foreign belief system to me, but again, as with Judaism, it was simply because of limited chances of exposure in my small town. After all of our early anti-Jewish Easter sermons gave way to the new "Israel and the Jews are our friends" teachings of televangelical preachers in the 1980s, Muslims became the new villains in fundamentalist evangelical churches. This religion was now a threat to Judaism and Christianity.

(After reading George Orwell's classic *1984*, I would forever compare my church's teachings on Jews to the classic "Oceania has always been at war with Eastasia" line. Depending on where you're at in the book, the two sides have always been friends, or always enemies.)

Any press coverage I'd seen regarding Islam in the 1970s and 1980s was always negative. Here was a chance to learn something new. I can't possibly do justice to a newly introduced religion in a partial chapter of one book, and I confess this is one of the least studied of the religions I'd encountered so far in my life, but I'll do my best to describe what I learned both in class and in the following years. It was interesting that a large percentage of the planet believed that a new prophet arose after Jesus. His name was Mohammed. This man, like Jesus, had what I would argue were good teachings, especially concerning charity and treatment of the poor and hungry.

Of course, there were new rules to follow. I instinctually rebelled against the requirement to pray five times a day. As I grew further from the dogma of my fundamentalist teachings, I had already ceased praying every single night. I'd already decided I was free to talk to God whenever I felt the need.

Our class seemed to react more critically toward this religion; perhaps those asking the questions were influenced by the same negative reports we saw in western media. There seemed to be a lot of violence, a lot of killing, directed toward non-believers (especially, of course, toward Jews).

An explanation was given for this that I could easily understand, based on my own upbringing: many religions had fundamentalist factions, and the ability to interpret scripture in ways that justified prejudice, mistreatment, and violence, wasn't just limited to Christianity. We should not attempt to paint Islam with any broader a brush than any other religion.

Conservative fundamentalism is not unique to single religion.

Another relevant question came, not surprisingly, from the women in our class: why did Muslim women seem to be treated differently than men? Did they not have equal rights and, if not, why?

There was a lot of handwaving over this question and, once again, it resonated with my fundamentalist evangelical teachings. Women were highly revered, and *of course* they were equal—but just in different ways. This reflected both my Protestant teachings, which said women should defer to men (especially in religious matters), but also to my Catholic upbringing, where women were good enough to serve in the church as nuns, but could never aspire to be priests. Because of the presence of strong, independent women in my life from the day I was born, I had rejected teachings like this from the moment I'd first heard them.

Islam had something else in common with my own religion: apparently nobody could agree on anything, and schisms had formed. We learned about the Sunnis and Shia; why those divisions occurred and how, similar to how my protestant congregation back home disavowed Catholics as true Christians, Islam had its own such divisions.

I admittedly had only a very brief introduction to Islam. I'd experience it on a much more intimate basis decades later, as I began traveling the world and spent time in Muslim countries. If not for the way people dressed, and the ear-splitting call to morning prayer from mosques near some of my lodgings, I could have been anywhere on the planet. People in big cities disregarded me completely, just as they would on the streets of New York. In small towns, I was warmly welcomed as a stranger and invited to more impromptu tea and sweets sessions than I could count. Most interesting though was that unlike in the United States, not a single person tried to convert me. I was never even asked about my religious beliefs.

The one negative, pervasive memory, which, if I'm going to be honest, I must report, was the treatment of women wherever I went. From large, cosmopolitan cities such as Istanbul, to small roadside, mud-hut villages across Niger, women seemed to be… not quite so equal as men. Ataturk had worked hard to establish a secular society in Turkey, but it didn't seem to be working. One hot summer day, with temperatures hovering around 95 degrees Fahrenheit, I sat in a commercial airliner on the tarmac of Istanbul's main airport.

The flight was delayed, and there was no air conditioning on the plane. As we sat there for an hour, I couldn't help noticing the passengers around me. The heat was stifling, even to me, dressed in Western-style cargo shorts and a breathable cotton polo shirt. All the women on the plane, approximately fifty percent of the passengers, were, without exception, wearing not only long black formless robes and hijabs that covered their head and neck but also, inexplicably (to a Westerner like me), ankle-length London Fog-style outer coats. If it was 95 on the tarmac, what was the temperature inside this jet? Decorum prevented me from asking, but the women looked absolutely miserable.

Just like the Catholic priest before him, our Islamic studies teacher could not give a satisfying answer to why women were equal, but not really equal. In my own personal views, which I realize may not mean much to the outside world, there should have been female priests. The women on that flight from Istanbul should have been free to dress in comfortable clothing. And yet, they weren't.

The cause for the same negative feelings I had toward fundamentalist Islamic and Christian teachings on women could also be found in Orthodox Judaism. There could be no female rabbis.

Having an entire university library at my disposal, and many history classes to come, I was free to dig deeply into the history of the Abrahamic religions. I'd come to discover that they were all established by men; scripture had been dictated and written down by the males.

It was a man's world.

Priestly Advice

The priest who taught one of my religion classes seemed to me a very knowledgeable, approachable man, and when he invited those of us with further questions to pay him an office visit, I took him up on it. I wanted to clarify a few points from his lectures, as he touched on something that had worried me from a relatively young age. The priest, mostly in an effort to explain why the Church opposed abortion, mentioned that we had an all-knowing God, one who, for example, "knew us from the womb."

Throughout my journey in religion, I had been bothered by some of the acts attributed to an omniscient creator. For example, why go to all the trouble of creating a world that you knew you were later going to be forced to destroy (via a flood)? If you knew the outcome ahead of time, why not skip the middle step?

But my question for the kindly priest was on an even deeper theological concept. As with all the questions I'd been asking religious teachers in my life, this was in no way intended to be a "gotcha" question. I wasn't trying to disprove the Bible; I wasn't trying to cast doubt. I had a serious theological question, and I felt like I'd found a reasonable, approachable man who might provide insight. Our conversation went something like this:

I led with, "So you mentioned in class that God is all-knowing. He knew us in the womb; he could number the hairs on our head. Would this mean that he knows everything about us, including our entire lives before we're born?"

"Well, yes, that would be a fair statement," he agreed.

"So he already knows the outcome of my life? How I'll end up in the end?"

"In a manner of speaking, yes, that would be true. You were created with free will; that's an important part of His plan. But yes, being all-knowing, he does know your fate."

Now, I had already been struggling with the concept of "free will." As I thought about the choices presented to me by my fundamentalist church, there did not seem to be a lot of freedom involved in the false dichotomy of "love and obey me or suffer eternal torment." But that wasn't the struggle that brought me to this kind man's office.

"So," I continued, "if God already knows all our outcomes, he's seen in advance how we'll exercise that free will. Since Hell exists, and we know people will suffer there, he will have accepted that many he created just aren't going to make it to Heaven."

The priest looked troubled. I don't think he expected the conversation to take this turn.

"I suppose you could look at it that way, but, again, he did give us free will."

Free will again. "I understand, and I do appreciate that, but being omniscient, he knew ahead of time how people would make their choice. So, I mean no disrespect, but isn't God creating people he already knows he'll be sending to eternal torment? Why would he do this?"

I felt bad asking this question. I can see how this might have appeared to be a theological trap, but that truly was not my intent. I did not want to put this kind man, who had invited curious students to seek him out, in the position of feeling I was there to mock or throw out religious paradoxes. Inspired by my ability to think freely in college, a freedom I never had as a fundamentalist, I was just looking for answers.

The conversation turned away from logic at that point. I heard many heartfelt words describing God's grace and his earnest hope that we would all accept his invitation, but no true answer to my paradox. I've never received an answer to this question that satisfies the logical problem it presents. I was very grateful that the priest took the time to talk with me. We had a very cordial conversation and, in the end, he left me with the same words I'd heard so often when I presented conundrums to my fundamentalist teachers earlier in life, words that would leave me just as troubled as when I'd struck up the conversations.

"Pray on this, Mark. Pray on it and ask for God's guidance. I know he'll give you an answer."

No answer ever came.

IDEX II

The year was 1985; my college days were coming to an end. At the time, I thought that these were the most enlightening days of my life, but I was wrong. More was yet to come. I did the rounds at job fairs, handing out resumes and talking to an endless stream of recruiters. The one job I really wanted, working for the now-defunct Digital Equipment Corporation (DEC), never came to be. I didn't even make it to the second round of interviews at the job fair in New Orleans. I was crushed. At the suggestion of friends, I visited the Lockheed Missiles & Space Co. table, dropped off a resume, half-heartedly chatted about my experience, and then went out to drown my sorrows on Bourbon Street. (The job fair just so happened to coincide with another college Spring Break. There were a lot of hung-over job candidates at this event.)

A few weeks later, back in Kentucky, the phone rang in my dorm room. It was a recruiter from Lockheed. He had a few exciting projects that I seemed to be a great fit for, and there were three separate hiring managers interested in talking to me. Would I be interested in flying out for a site visit and an interview? He apologized; the timing wasn't going to be quite right to watch the rollout of the new Hubble Space Telescope on its way to the launch site, but there were still a lot of interesting things he could show me.

Hubble? Space telescopes and satellites? Good Lord. I should have paid more attention in that New Orleans interview. Perhaps DEC turning me down was the gift horse I was looking in the mouth.

I flew to California and spent a day or two visiting the sprawling Lockheed complex in Sunnyvale. It was like a small city; it even had its own fire department. The interviews went well, and I ended up in the very fortunate position of getting offers from all three managers. The one hitch was that nobody could tell me exactly what I'd be working on. Lockheed did a lot of defense work; these were the years of Ronald Reagan's "Star Wars" programs.

Everything I'd be working on was classified, and I'd have to earn a security clearance to really know what the job was. Yes, the work was described in enough detail to sketch in some rough ideas, but the details would have to wait. I picked a project on the basis I'd be doing the very low-level programming I enjoyed, in a language known as "assembly language." It was about as close as you could get to talking to a computer without actually toggling dip switches.

It took eleven months to earn all my security clearances. Apparently, having relatives who still lived in Poland, then under the control of the Soviet Union, held things up. Also, during my polygraph tests, the interrogators wouldn't believe that I'd never tried illegal drugs. In the era in which I grew up, *everyone* had at least tried marijuana. Everyone but me. My Christian upbringing wouldn't allow it.

The polygraph administrator was persistent. "Look, Mark, we know everyone has tried pot. You're not going to lose your clearance for admitting it. The purpose of these questions isn't to trap you. It's to discover things in your past that might be secrets that could be used to blackmail you. Just get it out in the open and let's move on."

But I hadn't tried weed. I held my ground. If they wanted truth, I was going to give them the truth. I was sent away, then called back in later for a second round of polygraphs. I held firm. I eventually got my clearances. There was a plain old vanilla DoD Secret clearance, but also the crown jewel of an "SCI" (Special Compartmented Information) endorsement. I was allowed to see any information that pertained to my job, no matter how highly classified it was. And, oh, the information...!

It would be at least a decade before I was ever allowed to talk about what I did for a living. It was a strain to return to Kentucky for family visits and not be allowed to say a word about my work. "Come on, can't you tell us just a little bit?"

"No, sorry." I wasn't even allowed to utter in public the words that form the title of this chapter, the classified codename for our project.

Everyone loved to guess at what I did, often looking at my face to see if I'd give something away. It became a game we played on every visit home.

Finally, during the Clinton administration, IDEX II was declassified. I was allowed to talk.

Back in the early days of surveillance, actual rolls of film were loaded into satellites and blasted into space. Cameras would shoot photos as the "bird" flew over a target. At some point, machinery in the satellite would cut off the film strip, load it into a canister, and eject it into space, where it'd be pulled down to the planet's surface by gravity. A parachute would open and, *hopefully*, the canister would be captured by an airplane. By then, if the target in the photos was mobile, it had probably moved to a new location. Not a very efficient way of doing things.

The IDEX programs revolutionized the process. Film was replaced by digital cameras. Imagery could be beamed back to Earth in real time, and distributed to a large cluster of ground stations where analysts pored over the data. We could do color and higher resolution black and white. But the magic didn't stop there. By swinging the satellite into slightly different positions and gathering images from slightly different angles, three-dimensional images could be formed. Every workstation came with a crude set of goggles that had little windshield-wiper-like shutters on them. By flipping two images taken from different angles on a screen and synchronizing with the opening and closing of a shutter over the left or right eye, the brain was tricked into seeing things in three dimensions. My first such view was of Soviet nuclear bombers lined up on a tarmac. Scary!

For a recent college graduate who geeked out on technology, this was cutting edge. Our technology was almost all hand-built. There were no computer stores at the time, no Amazon, nowhere to run out and buy hardware. The motherboards for the ground station computers were etched out of copper-covered plastic plates in one of our workrooms.

Digital images are large, and at the time, we had no CD or DVD storage for imagery. The technology was on the way, but until it became widely available, we had to write software to emulate how a DVD "jukebox" would work.

Many years later, I'd stand in a hangar in the Smithsonian's National Air and Space Museum and look at the project that I'd played a very small part in. It was humbling to see something I'd spent all that time working on, now sitting in the shadow of another exhibit—one of the Space Shuttles. To be completely honest, our "advanced technology" looked rather clunky, plastic, and cheap compared to modern-day computers. But it was revolutionary for its time, and seeing it recognized as such gave me a sense of accomplishment.

But that sense of accomplishment is not what this chapter is about. I wanted to write about the *people* who made all of this possible. California was a true American melting pot. Whereas I grew up in a small, all-white town with only one religion, on the west coast I was introduced to myriad cultures, races, and faiths. Many of my IDEX II coworkers were from immigrant families just like me; some of them much more recent.

I made lifelong friends on this project. We were a tight group. Our rite of initiation was that we all started out as "mushrooms" as we waited for our coveted security clearances. "Mushrooms" because the company kept us in the dark and fed us s—t until we were cleared. We formed an ethnic food club; every Thursday we'd pick a restaurant either familiar to one of us because of our origins, or maybe from a culture that none of us knew. You have not lived until you've eaten Ethiopian food. Trust me on this.

I admit with shame that if our lunch group showed up at a restaurant in my small, conservative, religious Kentucky hometown, we would have most likely been eyed with suspicion, as outsiders. Especially if you had the wrong color skin.

More than once in my church, we discussed Genesis 9 and the Curse of Ham. Apparently, through some convoluted line of reasoning, because Ham discovered his father Noah drunk and passed out naked after the flood, Ham's descendants were cursed forever (why not curse Noah, the naked drunk?). Intolerant "scholars" concluded that "Ham" was closely related to the Hebrew word for "hot" to mean "black" or "charred" and therefore, people with dark skin were inferior and cursed. (Again, these are things I can't possibly make up. Those conservative fundamentalist White nationalists we see on television today? I have an idea where they got some of their information.)

I was told more than once in church that God just didn't intend for people of different races to intermix. I would often look at the painting of White Jesus hanging behind the altar in our church. I'd seen plenty of television from the Middle East. Shouldn't

Jesus have had dark skin?

It's hard to convey to modern-day Americans that these fundamentalist religious/political beliefs were not just the twisted thoughts of a small fringe group that existed fifty or sixty years ago. These ideas are very much alive in 2024 as I write this book.

Our amazing IDEX melting pot was, as the saying goes, as American as apple pie. There were at least four languages spoken on our project. There were many religions and people of all ethnic backgrounds. Many of us didn't look like what people in my small, conservative hometown would call an "American."

Yet there we were, Americans all, contributing to something our nation would showcase in its most famous museum as a national achievement.

Kalalau Trail

L iving and working in California had opened up a new world for the outdoorsman in me. I'd loved being out in nature since the very first camping, fishing, and hunting trips my father took me on. With California's Sierra Nevada range at my backdoor, I was now able to get into serious backpacking, spending a week or so in the wilderness with nothing to rely on but my wits and whatever I carried with me. Getting ever more adventurous, I learned cross-country skiing and traversed the upper reaches of Yosemite National Park by ski, sleeping in a tent pitched in the snow every night for a week. *Oh, the peace and quiet!*

The most challenging adventure, though, was the Kalalau Trail, on the Hawaiian island of Kauai. If you've seen movies such as *King Kong*, *South Pacific*, or *The Perfect Getaway*, then you know Kauai—soaring green, eroded volcanic cliffs surrounded by beautiful white sand beaches. Paradise on Earth.

The Kalalau Trail winds its way eleven miles along the northern end of the island, a region known as the Na Pali Coast. The trail is considered one of the most dangerous in the world and is not for beginners. Inexperienced hikers die here. Most mistakes occur when crossing raging streams, the victims swept out to sea. But the exposed, knife-edge cliffs that are a main feature of the path are what scare the willies out of people like me, who suffer from a bit of vertigo. In some places, the trail is no wider than the width of a human's shoulders, with a volcanic cliff soaring hundreds of feet into the sky on one side, and a drop of as much as 800 feet into the ocean on the other.

I did the trail for the first time in my mid-twenties. It was the experience of a lifetime. Except for the incessant buzzing of tourist helicopters, which would come whipping around cliff edges unheard and startle me to the point of almost falling over the cliff edge, I was largely alone in paradise.

Until *they* showed up.

The pair of hikers approaching me from the opposite direction looked very out of place. No backpacks, no proper hiking gear. Just two young men wearing aloha shirts, flower leis, and carrying small books in their hands. They smiled and greeted me.

Oh no.

Jehovah's Witnesses.

On the edge of a cliff in the middle of the wilderness?

How in the world? I'm used to knocks on the door as I'm climbing in the shower. I'm used to the doorbell ringing when I'm about to burn my bacon. But eight miles into the wilderness, on the edge of a cliff? I was trapped. I couldn't ignore them. There was no door to hide behind; I couldn't pretend I wasn't home. I couldn't turn around and run. I was carrying a seventy-pound pack and they had only Bibles. They would have caught me easily.

"Hello sir! Beautiful day! Do you have a moment to talk?"

They were blocking the trail. With no way forward, I sighed and dropped my pack. I needed some water anyway.

"Uhm, sure. What about?" As if I didn't know.

"Well, we're missionaries, and we're out here today to witness to the marijuana growers in the jungle at the end of the trail, and to anyone we might find along the way. Have you heard the good news about Jesus Christ?"

Why yes, in fact, I had! I screwed the top back on my water bottle, thanked them for bringing it up, and mentioned I was in a hurry. I still had three miles to go and the day was getting late. I hefted my backpack and looked at them expectantly. As in, please place your backs against the cliff so I can risk my life trying to squeeze around you on this knife-edge cliff.

They weren't giving up that easily.

"If you don't mind sir, it would mean a lot to us if we could just share a favorite bit of scripture with you and then we'll be on our way…" Without waiting for permission, one of them opened his Bible, walked over to me, balanced precariously on the edge of a straight drop into the ocean, and smiled while his friend took our picture. One slight bump from me and… no, I'm just kidding. I would never do anything like that. But it *would* have looked like an accident.

So I received my daily Bible verse, some unwanted hugs from the missionaries, and then they edged around me, headed back toward Hanalei to find their next victims. Somewhere out there in the world, sitting on a dusty shelf, is an old copy of Watchtower, with a photo of a disgruntled backpacker standing next to a missionary who could have *accidentally* gone over the cliff edge.

I was incredulous. Even in my younger, evangelical days, I would have never thought to have intruded on a stranger going about their business on a narrow volcanic path hundreds of feet above the Pacific Ocean. I was a devout, dedicated Christian back in my day. I was fervent in my beliefs. But there were (I thought) limits to everything.

All I wanted was to be left alone. That's a sentiment I hear a lot these days from atheists, agnostics, humanists, even members of other religions. Just please leave me be and let me go about my life in peace. I'll be the first person there to defend anyone's right to worship, but if they could just leave me out of it, yeah, that would be great.

And by the way, don't try to interject your beliefs into our government or our classrooms.

Is this too much to ask?

Close Shave

The melting pot of California was a life-changing experience for me, and the lessons I learned there will stay with me forever. But I am a family man. My heart and mind belonged back in Kentucky with my loved ones. So, after five great years in Silicon Valley, I tearfully said goodbye to my friends, packed up my belongings, and made the three-day cross-country drive to the Bluegrass State, where I'd landed a new job in pharmaceutical computing.

Was I ever in for a culture shock. It's amazing how just five short years can change a person.

One brief story, involving bicycling, will help me explain.

The 1990s and 2000s were a great time for cycling in both Kentucky and the United States, especially for those of us who were into racing. Lance Armstrong would pick up his first World Championship and go on to seven Tour de France wins. Closer to home, Lexington, Kentucky could still be called the Horse Racing Capital of the World—which meant countless miles of country backroads servicing horse farms, perfect for the 50-to-100-mile training rides necessary to compete even at the most basic amateur level.

Of course, Lance Armstrong would later go down in utter disgrace, exposed as a drug cheat. Lexington would continue on its inexplicable march toward plowing under every horse farm to build unwanted shopping malls and high-priced rental units.

But for a time, the backroads of Horse Country were my

playground. I shelled out everything I could spend on a proper racing bike. I bought a closetful of Spandex bibs and shorts, and colorful cycling jerseys that would put Hawaiian shirts to shame in a color contest. I learned to ride with special pedals that accepted a cleated cycling shoe, so that I could "clip in" like a downhill ski racer, firmly fixed to my bike to increase pedaling efficiency. I would ride 6,000 miles a year or more, in all weather conditions, just for the chance at the occasional, elusive win in a mass start road race or criterium.

Criterium racing was the ultimate adrenaline rush for a young man or woman who still thought they would live forever. Taking place on shut-down city streets, criteriums were akin to the Formula One racing you see on European television—fifty or more brightly colored racers going top speed around the tight corners of city blocks, riders sometimes separated by centimeters, often actually touching each other, shoulder slamming into shoulder as everyone angled for the best line through a corner. In a "crit," my speedometer would start off around 26MPH and only climb higher as the thirty-to-forty-five-minute race progressed.

And, oh yes, I shaved my legs.

Suffice to say, men shaving their legs in the Bible Belt was not something that was commonly seen. Contrary to popular belief, cyclists do not shave their legs in order to gain a speed advantage. We did it because of the crashes. The wrecks were savage. If you were lucky, the worst thing that happened was a lot of lost skin. In worse cases, bicycles would be broken into pieces along with the bones of riders.

But the shaved skin? It was to protect against infection. You don't want debris in a wound. Especially not tiny leg hairs.

I once went down in a corner on a freshly paved city street. The asphalt had not quite cured yet, and was a bit slippery in the hot summer sun. The little black pebbles that are inevitably left behind from a fresh paving job didn't help. One moment I was traveling into the corner at a speed with which I knew better than to try to take corners (I was trying to break away from the pack and win). The next few moments consisted of my tires losing traction, the bike flying out from under me, and my body sliding across the entire width of the road at around 30MPH.

It didn't hurt at first. I stood up and looked in amazement. There was a long streak of blood and shredded Spandex ground into the asphalt, leaving a trail all the way from the corner to where I stood. I looked down at my leg. The clothes on the right side of my body were gone. My leg, from my ankle to my upper thigh, was bright red. The pain hit me then. I said words I won't repeat here.

Later, in the emergency room, the nurse brought out a bottle of Phisoderm soap and a scrub pad you'd see someone use when washing dishes.

"This is going to hurt," she said. "Good thing you shave your legs. It's going to be hard enough to get the rock and dirt out. If you had hair in there, this would be much worse." Well, at least one Kentuckian understood my leg-shaving. I was very soon to encounter one who did not, however.

The nurse started to scrub. The injury wasn't serious enough for pain medicine. But every piece of debris was an infection risk. Before they bandaged me up, they had to make sure the wound was clean. Remember, my skin had been ground to something resembling raw hamburger. I'm not a baby, but I couldn't help letting out the occasional yelp. On and on she scrubbed.

"Try and hold still," she patiently said.

Shaving my legs seemed like a wise precaution, and this experience confirmed it. Both I and the nurse were glad I had baby-smooth skin, with no hair to cause infection.

Others weren't so appreciative.

One evening, after finishing a group ride as darkness fell in a small suburb of Lexington, I tied my bike to the rack on the back of my car and began the drive home. My fuel warning light came on. The needle was sitting on "empty." I didn't want to risk it, so I pulled into the first gas station I could find. I hadn't bothered to change out of my cycling clothes.

Spandex and shaved legs, standing at a gas pump in a very rural, conservative town. This could not possibly have ended well.

I heard a loud voice from the car on the pump opposite me. "Look at that f****t. Son, that's what a q***r looks like."

What?

I looked into the car and saw a man about my age sitting behind the wheel. A young boy was sitting in the passenger seat.

"What did you just say?" was all I could come up with.

"You heard me, f****t. What kind of man shaves his legs and dresses like a woman?"

Reasoning with people like this was pointless, but I tried: "I'm a bicycle racer. There's a reason we dress like this…"

I was interrupted with a stream of cussing and threats that, again, don't need repeating here.

All I could think of was, "What the Hell are you teaching your son?"

"I'm teaching him about life. I'm teaching him about God. I'm teaching him what the Bible says is wrong with people like you."

And there it was. God. The Bible. People like me.

Just like I'd been taught in church when I was little. Exactly what I heard from evangelical friends, some of whom I attended Sunday school with, on school playgrounds. This was still being taught in churches. It's still being taught as I write this book. Looking at the little boy in the front seat of that car, I had little hope that he would turn out to be unlike anything other than the man sitting behind the wheel, whistling and catcalling as I tried to control my temper and walk away. Never mind the bigoted driver; unless logic and reason intervened in his little boy's life, there was a carbon copy of his dad, already in training.

My girlfriend and I would later laugh at the irony of this event, after one of those special intimate moments that adults share.

"Well, you're still not gay," she'd laugh.

Odd thing is, the man doing the name-calling somehow thought that he was insulting me by assuming my orientation. Having loved ones—friends and family—who were LGBTQ, had long since moved me beyond the point of considering being gay as "wrong," even though that was drilled into me in church. But the way he did it was deeply offensive. I felt offended for my friends and family. I cannot speak for them—I have not lived their lives and do not have the experiences they've had—but I have listened to them, and they assure me that these types of incidents are common.

The incident at the gas pump was an unwelcome reminder of what evangelical churches taught me. A reminder of what some are still teaching. A look into the abuse some members of our society go through in their daily lives. A very ugly reminder.

College Years (Again)

I was lucky enough to begin my professional career in aerospace, a field that fascinated me because of my background in electronics, computers, and astronomy. While I can say that my favorite professional project will always be my IDEX II years at Lockheed, the most inspirational work I'd do would actually be in the medical and pharmaceutical fields.

A friend had recommended a pharmaceutical software company in Lexington and, after a short interview, it turned out I was a good fit for the job. The ultra-conservative corporate culture in Kentucky was a huge, negative shock to me after being in diverse and free-thinking California, but there were trade-offs. I was getting my hands wet in another branch of science that fascinated me. Chemistry was one of my favorite subjects in school. Now they were going to pay me to do it?

I took over as the lead programmer in my new company's hospital pharmacy division. One of my job duties was to spend time in hospitals, training with and going on rounds with pharmacists. Being a software developer is somewhat akin to being a translator of languages. I not only had to understand how to program a computer, I had to understand pharmacy at least at a level where I could translate what a fully trained pharmacist was telling me, into tasks that a computer could perform.

This was a job I took very seriously. Yes, our company had a ton of lawyers warning that the hospitals were responsible for checking all of the work that we generated, but lives were literally still on the line if I made programming errors. My morals ran deeply enough that if a mistake on my part harmed another human, it wasn't going to help my emotional well-being at all to have an attorney whispering in my ear, "Hey, they signed a disclaimer. You're in the clear."

In holding my job so closely to heart, my interest in chemistry and biology deepened. These were two topics that I'd barely been able to touch upon while earning my four-year bachelor's degree in computer science. There were only so many science electives I could cram into those years. Because of my AP chemistry in high school, even though I'd blown up a moonshine still as part of my experiments, I tested out of college chemistry. Biology was limited to a single general course that filled in some blanks, but never delved into the depth of detail necessary for understanding topics like evolution to the extent I desired. It wasn't the instructor's fault— there was just too much to cover. We expanded on some of the evolutionary basics I'd learned from Sagan's series *Cosmos*, but just barely.

So, my curiosity and love for my work eventually landed me back in college, this time as a pre-med student. Approaching the old, old age of thirty, I'd decided I wanted to switch careers and become a doctor. This wasn't an easy sale to my employer. I was still required to get in my forty-to-fifty hours a week, make hospital visits, and also manage to squeeze in two years' worth of all the science courses I hadn't been able to get in during my first college stint: organic and inorganic chemistry, animal biology, human biology, microbiology, human physiology, medical ethics.

It was going to be a tough road. Thanks to night classes and an agreement with my company to start work at ungodly hours so I could take long lunches to attend classes only taught in the day, I enrolled as a student at a major university that featured a leading medical teaching hospital. It didn't occur to me until I stepped into my first classroom that I was at least a decade older than my classmates. My fellow students were actually pretty cool about it.

After a few strange looks, once I'd befriended those around me and they found out why I was there, I was fondly accepted as "the old guy."

Gaps in my limited knowledge of evolution were filled in almost immediately. Fossil evidence. Homologous anatomical structures. The undeniable evidence presented by DNA. The gorilla that had so fascinated me during a visit to the Milwaukee Zoo decades earlier? Approximately only two percent of our DNA differed.

Just two percent.

Back in that zoo, my hunch that I was looking at a cousin was confirmed, not by just one representation of a family tree, but two separate ones. There was a "phylogenetic" tree based on DNA. There was also a "morphological" tree, based on physical characteristics such as skeletal and anatomical similarities.

A simple thought experiment was given to the class: the fossil record clearly showed that nearly all life that had ever existed over billions of years on Earth was now extinct. Countless species—the majority of species—were all gone. Gone forever. Modern species, like we humans, had only been around, at best, for a couple hundred thousand years. How did we get here? Where did we come from? Did we magically appear out of the dirt, or did we have ancestors that no longer lived?

I looked at the family trees. The answer was obvious.

We did discuss so-called "living fossils," such as a fish known as a coelacanth. This fish had apparently been around in, more or less, the same form for over 300 million years. It was known from the fossil record and, thought to be extinct, was rediscovered, alive, in the 1930s. Even with my limited evolutionary knowledge, I correctly guessed the answer to the riddle of why this creature hadn't evolved—It simply didn't need to. It was under no pressure to do so.

A key driving force behind evolution is natural selection: as the environment changes, creatures who evolve to be a better fit for that environment tend to survive and pass along their genes. Coelacanths lived in caves very deep in the oceans. There's not a lot of change going on down there. Where was the pressure to evolve? It wasn't there.

It isn't my intent to try to teach a course on evolution in this book. I don't have the expertise for that. What I'm attempting here is to highlight some of the very basic things that I could have—should have—been taught on this subject much earlier in school.

I frequently encounter fundamentalists who want to debate topics such as evolution, and I quickly learn they've never studied it anywhere other than Facebook. I'd be rich if I had a dollar for every time I've heard "Evolution is just a theory." Well, gravity is "just" a theory. The problem, I think, is a lack of teaching of basic science fundamentals.

Inevitably, evolution deniers don't understand that a theory is the highest ideal in science. They use the colloquial "just an idea" definition instead. And coelacanths not evolving? You'll get that challenge in every debate you have with an evangelical who's never studied evolution. Bet on it.

Elsewhere, in various courses, we touched upon information that would profoundly impact my views on a topic that has been controversial since I've been alive:

Abortion.

My religion had taught me that all life begins at the moment of conception (though they were too shy to explain what "conception" meant). I was to learn more nuanced views through science. A scientist might argue that a cell is "alive" because it has certain characteristics such as a metabolism, DNA and other biological structures, and can replicate itself. But science never taught me that a gamete (the result of a sperm cell merging with an egg cell) was "alive" in the context of being a person. It was, quite literally, a single cell that, under the right conditions, might go on to produce a human. Or maybe not.

I did learn that, once again, I wasn't being given the whole story, not just by politicians, but even (unintentionally, I think in this latter case) by some of the medical community.

My fundamentalist church often told us that one of the many ways we knew there was a live baby inside the mother is that a heartbeat could be heard just a few weeks into the pregnancy. When I was young, I accepted this without question. I was wrong. My church was wrong.

The revelation came during a section on embryology. We learned that in a human embryo, the heart, as we typically know it, hasn't fully formed that early in pregnancy. At six weeks, for example, the "heart" is a simple straight tube, surrounded by nascent cells differentiating into heart cells. The tube begins curling back upon itself around this time, forming loops and other structures that eventually become the heart we all recognize.

The four-chamber organ with the familiar "lub-dub" sound we're familiar with (the sound being made as valves open and close)—hasn't developed. And yet, I vividly recall the excited, expecting mothers at church confirming they'd heard the baby's heart beating for the first time.

The misunderstanding seems to have been perpetuated not only by the internet (that fount of all knowledge), but also by some medical clinics who were making an innocent but, as we shall see, perhaps dangerous intimation that a fully formed human heart was at work. This is all due to the device being used when the mother visits the clinic: a doppler ultrasound.

This device works by producing sound waves which travel into the mother's womb, reflect off tissue such as the developing heart, and are then reflected back to the machine. The velocity of sound waves is altered as they strike tissue inside the womb. The machine *artificially* produces a sound based on the differences between the sound waves emitted and those received when they bounce off fetal tissue. This is where the "heartbeat" sound comes from: it's produced by an electronic device. Unplug the machine, remove the sound. No heartbeat could ever be heard by a stethoscope at this stage.

Later in my life, the American College of Obstetricians and Gynecologists would begin pointing this fact out to clinics who invited mothers to "hear your baby's heartbeat," but it doesn't seem everyone is listening to the science. Lawmakers certainly aren't. Just a few decades after I learned these facts about embryology and the development of the heart, politicians would begin introducing so-called "heartbeat" laws that would prohibit abortion after six weeks.

I felt that people were being deceived. Just as my church had insisted decades earlier, legislators seemed to be equating the development of a six-week-old embryo with a fully formed person. You could hear the heart beating, after all. Right? This would seem to me to be an emotional appeal, not one based in science.

One of the major reasons I so loved science education was on full display here. I would argue that in the technological times in which we now live, the better educated we are, the better able we are to address the complex subjects being thrown at us on a daily basis. While I freely admit that I am not a doctor or a professional scientist, I feel that my secular education did indeed give me a solid basis upon which to make informed decisions on topics that are hot-button issues in our lives: abortion, climate change, vaccines, and more. I'm loath to turn to social media (as so many seem to do) for answers.

Back to abortion... there were many other pieces of information I gleaned that would dissuade me from later automatically saying, "Oh, that's a baby," as my church would have me do. Fundamentalism taught me that it was a baby from the moment of conception. An embryo, to my congregation, was a child.

It was fascinating to me that science was developing the ability to freeze an early embryo to minus 300 degrees Fahrenheit, then later thaw it, implant it, and let it grow into a baby.

You can't, under any circumstances, freeze a person to 300 below zero. It just doesn't work. Again, I must beat the proverbial dead horse here—this information was presented to me as a simple fact—actually a wonderful scientific achievement—not as an argument in favor of abortion, a subject that was never mentioned.

Fantastically, as this book goes to print in February, 2024, Alabama's Supreme Court has just ruled that a frozen embryo is in fact a human child, and in-vitro fertilization clinics are now worried they could be charged with murder if they destroy an embryo. No scientific explanation was given by the court in this decision. In fact, the judge who wrote for the majority made references to God and religion to justify the result.

I mention these facts not as arguments in favor of abortion, but rather as examples of how one side of the argument might be using scientifically inaccurate and/or religious information to emotionally sway us toward a certain conclusion. Uninformed decisions can cause enormous problems for doctors and scientists who are trying to help. For example, as we learned in pre-med, it's standard to produce a large number of embryos to increase the chances of a successful pregnancy in IVF treatment.

Now, clinics must grapple with dilemmas such as what to do with remaining embryos should they achieve pregnancy early on in the process. Or, perhaps a couple gets divorced, or for some other reason decides not to have children. What happens to the embryos? Must they be kept frozen forever? Will doctors or hospitals be charged with murder if they destroy them? This court decision has only just now come out, but already pundits are speculating this could be the end of IVF treatment in Alabama, shattering the dreams of parents who could not otherwise conceive.

Perhaps the most impactful teaching on the delicate, controversial subject of abortion came not from natural science classes, but from a required course on medical ethics. More than any other class I took as a pre-med student, this course forced me to confront head-on thoughts that I, and I suspect many others, have simply pushed out of their heads with a quick "no, that's wrong," or, "yes, that's OK."

Medical ethics is the one class where I wasn't just asked to form an opinion. I was *required* to form one. Or, more accurately, critically examine any opinions I might already have.

The reasoning was simple: Every student in that classroom would inevitably be forced to make medical decisions in situations they may have never considered. Or, perhaps more importantly, they *had* considered subjects and had, perhaps unknowingly, arrived at a position that did not meet the all-important purpose of providing the best care possible to a patient. And, we were told, providing the best possible care meant honoring the needs of the person we'd be caring for. Our personal thoughts on a matter were secondary.

We didn't start with the topic of abortion. Assisted suicide was first on the list. Had you questioned me going in, I would have been of the same mind as any of my former church members: it's a living person, you have no right to intervene. God makes those decisions. Period. After looking at case studies of humans in agony while dying from cancer from which there was no hope of recovery, I changed my mind. And, there were those who were technically brain dead, but were left hooked up to machines to keep them "alive" when there was no hope of recovery. I came to see this less as medical care, and more as cruelty.

On the subject of abortion, some common myths, all which I'd heard espoused by my former church, were dispelled. Few women were using abortion as "birth control." My preacher told me that these were just loose, irresponsible females who should have known better and, I quote, "kept their legs closed." Our class discussed rape victims, some of them actual children themselves, often impregnated by family members. How did we feel, our instructor asked, about forcing kids to have babies?

We talked about myriad cases where a pregnancy goes horribly wrong and there's no chance a living child will result. We spoke of horrific cases of fetuses without fully developed heads, or missing organs such as kidneys. There were pregnant mothers who discovered they had aggressive cancers and therefore needed to begin chemo or radiation immediately. Time was a major factor in treating cancer. Yet these mothers would be forced to forego treatment until they delivered—they couldn't be treated while they were pregnant.

Without knowing it at the time, we were speaking of a friend I'd make later in life, who was told, with no wiggle room, that due to a significant complication, if she continued her non-viable pregnancy, it would end her life. There was no chance of a live birth. What was she to do? She was a strong Christian. She already had a young son. Who would care for him if she continued the pregnancy and died? Listening to this friend speak, it was clear that this was the most heart-wrenching decision of her life.

In short, the subject of abortion was far more heavily nuanced than my fundamentalist brethren had led me to believe. The choices were not always black and white. Much of what I'd been taught as a youngster did not have a solid basis in science. And to say empathy was lacking for those who might become pregnant would be an understatement.

In medical ethics, we talked about situations where birth control could fail (as it does one or two percent of the time), and responsible adults who were taking every possible precaution ended up pregnant anyway. We read testimonials from single women who lived in poverty, who already had children and couldn't afford food, let alone day care, and were forced to leave their children alone so they could go out and work multiple jobs, for minimum wage. As I've said elsewhere, but it bears repeating: Fundamentalists sometimes openly called such women "welfare mothers," and their offspring were "welfare babies."

Once you made it down that birth canal, at least among my fellow church members, you went from being a blessing to a burden on the system. They would insist that the baby be born regardless of any and all complications, then, in what seemed to me to be a great act of hypocrisy, label the new infant a "welfare baby," condemn the mother as a loose and immoral person, and in some cases, actually openly campaign to deny the mother and her new family social benefits that would help them survive.

Of course, not all Christians see things this way; many work to help children. But my fundamentalist church would not agree with this. We were firmly opposed to simple, effective ways to prevent unwanted pregnancies, such as sex education and birth control. As with all unyielding conservative viewpoints, the subject was black and white.

And so, on my final exam in medical ethics, one of the questions was to either defend or oppose not abortion, but a woman's right to make her own choice, based on her own particular life circumstances. I paused on this question, pen in hand, poised over the paper. I loved children, and hoped to one day have many of my own. But, based on rational thought, I had to admit there were many instances where a woman would legitimately seek this type of medical care.

Further, I had to concede that, especially being a male, I was being asked if I should have the power to tell another human, a female, what to do with their body. I could not do so.

I was already determined that, as an adult, I was going to make it very clear that the choice for my own body was that I'd never consent to be put on a life support machine and left in a vegetative state for years or decades. I'd make that choice in advance, and make it legal. I'd make the decision for myself. Nobody had the right to make it for me.

How could I make a medical decision, of any type, for someone else?

College did not make me "pro-abortion," as the slur is often used by the religious right. It made me, in the very literal sense, pro-choice. I am not a woman and therefore not qualified to speak for them, but I do know those who've had to face the agonizing choice we're talking about here, and hearing their words, I can understand: this is perhaps the most difficult choice anyone will ever face in their life. It is excruciatingly painful to listen to the stories, put yourself in their shoes, and not come to an understanding different than what I was taught in a fundamentalist, Young Earth Creationist church.

I elected the "pro-choice" position on my final exam, and defended it vigorously. The tug of my evangelical past surprised me by resurfacing as I wrote. How could I betray my church like this? Was I now a "baby murderer?" My rational mind kicked back in, though. I was operating on knowledge that was never made available to me during my fundamentalist training. Black and white, "one size fits all" positions were no longer a part of my life.

I've spent a lot of time in this chapter discussing, more or less, a single topic. Perhaps this is because it has become such a major point of debate in the USA as I write this book. But abortion as health care is not the only meaningful lesson I would take away from my second stint in college.

My first four years of higher education laid the groundwork in the natural sciences I would need to understand that we had in fact evolved as a species; to understand the true age of the Earth and the universe; to be able to give informed replies when later confronted with unscientific claims presented by organizations such as Answers in Genesis (AIG) and the Institute for Creation Research (ICR).

As a former fundamentalist, I can more easily understand, and perhaps sympathize with, those who fall for the traps set by shiny creationist theme parks that falsely claim that humans walked with dinosaurs.

Now that I had a solid core in biology and microbiology, I could see how someone might deny vaccines, given that they weren't fortunate enough to have had a chance at college. It would be arrogant of me to say that everyone needs a university degree to have an opinion on such subjects. And yet, it is clear to me that a great deal of misunderstanding in today's world seems to come from "education" gleaned from YouTube and Facebook rather than scientific sources.

Among those non-scientific sources would also be AIG and ICR. I can say, and back with evidence, that these groups are not teaching science. They're teaching science denial. Yet, wrapped in the cloak of the word of God, they are very hard to refute and debunk. As was taught when I was a fundamentalist, any disagreement with such institutions and their teachings is to literally argue with God, and we all know what happens to people who do that.

In my childhood, I was literally threatened with Hell because I believed that, based on physical laws, rainbows had to exist as soon as sunlight and raindrops were present. But I was, according to my fundamentalist teachings, openly disagreeing with God. Speaking from experience, this is a very hard trap from which to break free. It's very difficult to disagree with a supreme, divine being, or even someone who claims to be speaking for that divine being.

The science denial I witnessed early in my life was very clearly born from fundamentalist teachings. This dark cloud of denial seems to be spreading through schools as I write this book. Although I can't think of a better defense against misinformation than science and logic, those very topics are being slowly erased from public education by, you guessed it, fundamentalist, conservative religion. Having studied, for example, evolution, I can see the truth in it, and understand why it's so important to organizations like my childhood church to keep it out of school curricula. If you don't want students to consider something, simply prevent them from seeing it.

I know many, many Christians (and members of other faiths) who understand and accept evolution, and it has not tainted their beliefs in the slightest. In my experience, when you discover that your respected elders did you the disservice of hiding things from you, it pushes you further away.

You eventually stop believing what they're telling you.

Heart Surgery

Midway through my pre-med studies, I was befriended by a doctor in our local cycling club. Dr. "S" was a true free spirit, as unfiltered as they came. She cursed like a sailor and told dirty jokes to the peloton on long rides. We got along really well. She was an anesthesiologist and when she found out how interested I was in medicine, surprised me by asking if I'd like to observe a couple of open-heart surgeries. I thought she meant something like on television, where people sit in a room looking through glass onto an operating theater below.

No, she meant *observe*. As in, scrubbing in with the rest of the surgical team and being in the room. Not just being in the room, I would discover, but actually taking her place on a short stool at the patient's head, looking down into an open human chest.

It was beyond my imagination that I had enough "street cred" to be considered for an opportunity like this. I can only guess at the combination of factors that gave me this chance. I'd been doing rounds with pharmacists in hospitals I wrote software for. But ultimately it had to be the fact that Dr. S. was well respected at a top medical teaching college, and she was friends with the head surgeon, who gave his OK.

Once again in my life, a teacher or mentor would recognize in me someone eager to learn, and step in to make that happen. I hope that I may be given the opportunity to pay this forward in the years I have left on Earth.

I took a vacation day at work and showed up at the hospital early. There were to be two surgeries that day. One was a "CABG," a coronary artery bypass graft. An artery in the patient's heart was obstructed, and blood flow needed to be rerouted. Two surgeons would work at once. One would open up the chest, the other would open one of the patient's legs. A vein would be removed from the leg, then passed to the heart surgeon, who would stitch it around the obstructed artery in the patient's heart, allowing blood flow to resume.

The second surgery of the day would be an aortic valve replacement. This one was a little more serious. An important valve in the patient's heart wasn't functioning properly. There were two choices for a replacement—a metal tube, or the organic route—a valve from a pig's heart that had been stripped down to cartilage and removed of any material that could cause the patient's immune system to reject the valve.

I scrubbed in under the very (very) watchful eyes of the head nurse. Those quick scrub-ins you see on TV doctor shows? Forget that. This was a thorough, and somewhat torturous experience for my tender skin.

"You missed a spot."

"You missed that spot. Scrub harder over there. Scrub!"

Finally, skin red and tingling, I met the nurse's approval. I held my arms out, was wrapped in a surgical gown, masked up, and ushered into the operating room. The first thing that hit me was how relaxed everyone was. This was routine business to them. Dr. S. showed me to her normal station at the head of the table and helped me step up onto a footstool. Her last words of wisdom to me were:

"Not everyone can take this. If you have to puke, turn your head to the side. Don't vomit into the patient."

What an amazing moment. The surgeon greeted me, asked me a few questions, and then asked for music. Someone turned on a cassette player (once again, yes, I'm that old). Pat Benatar blasted out of the speakers. I'd idolized Benatar since buying her first album in high school. Wow. Rock and Roll and surgery.

The first operation was pretty routine, if someone with absolutely no surgical experience is worthy of saying such a thing. The patient's chest was opened before my eyes, bones spread apart by metal devices to give the surgeon easy access. The knives being used automatically cauterized blood vessels as they cut, to minimize bleeding. There was amazingly little blood, but the sickly-sweet smell of burning flesh quickly took over the room. Nobody seemed to notice.

I quickly got over it. I was looking down on an actual live, beating human heart. People were afraid I was going to get sick? Seriously? I was living in a dream. There was so much to take in.

The patient's left leg was opened by another doctor. A vein was removed and passed to the head surgeon in a sterile metal bowl. While the leg was being closed up, he went to work. Along the surface of the heart, it looked like he made small incisions around the occluded area, then quickly stitched in the leg vein to bypass the problem. I could see blood begin to flow through the vein. To everyone in the room but me, this was routine. To me, it was magic.

It seemed to be over before it began. The patient's chest was closed. I looked up at the clock and was surprised. What seemed like minutes had actually been hours. It was already time for lunch. Good. I was starved.

Chinese take-out was on the menu, delivered to a room just

outside the OR by an orderly. I dug in. Dr. S. was in good humor and confessed she was a little surprised. "Even some of the residents fail to get through that without puking. We had one poor guy who fainted. Luckily, he fell off the stool backwards and didn't land on the patient."

I admit, I felt a little pride.

Lunch being over; it was back for the aortic valve replacement. But first: Scrub, scrub, scrub.

"You missed a spot."

"No, more time on that spot over there. Scrub!"

Everything went as smoothly as the first surgery. Until it didn't.

The patient was an elderly woman. She was placed on a bypass machine during the surgery. Her heart had to be stopped, and the machine took over for her heart, circulating blood, while the surgeon did his work. When he'd finished, she was taken off bypass, and two metal paddles were placed alongside the heart to give it the shock necessary to restart it.

Zap!

I watched in amazement as the once-still heart began beating again.

And then, it stopped.

Even to a medically illiterate computer scientist such as myself, things had clearly gone wrong. Fascinated by technology, I'd spent part of the surgeries watching the various computer displays and trying to interpret lines and numbers. I didn't need to be an expert to realize the screens were screaming out "problem!"

The patient's heart wasn't beating.

"Music off!" were the surgeon's first words. Pat Benatar faded to silence. The surgeon began reciting instructions. Dr. S. walked up to me, pointed to the floor, and said, "Down! Over there! Quickly!" I got out of the way.

The next minutes were a battle of life vs. death. Shocks. Vasodilators. Vasoconstrictors. A lot of cussing. The patient was dying. Or was she already dead?

"Come on, God d—mn it! Come on!" That was the surgeon speaking.

After what seemed like an eternity, the monitors I'd been watching came back to life. And so had the patient. Dr. S. moved away from the head of the table. Her face was pale and covered in sweat. She motioned me back to my perch. I looked down, and saw a beating heart.

Closing up was a solemn affair. I'm not a doctor and not qualified to speak on medical affairs, but the atmosphere in the room told me that this had been a close call.

"That was close. That was really close," Dr. S. confirmed when we finally sat down outside the operating room.

The takeaway moment from this day, the point of this story, was yet to come. Dr. S. and I walked to a small recovery room. We passed the surgeon, slumped in a chair. He was staring straight ahead, not looking at or talking to anyone. I wanted to say how much I admired him, but I couldn't find the courage to speak. What was he feeling? Could I ever in my life undergo as much pressure as he must have felt?

We walked into the recovery room and waited while the patient, an elderly grandmother, recovered from the anesthesia. She was surrounded by what I assumed to be her children.

"Thank you, God. Thank you, Jesus," the relatives said again and again.

The surgeon remained in his seat, staring at the wall, while praise was shouted to Heaven. This took me back to an earlier time when quick action by humans, paramedics, and doctors had saved the life of my cousin Michael, who'd been hit by a car. God got all the praise. The humans who took all the action were an afterthought.

I understood the family's relief, and I know that in trying, emotional times, people will express feelings in their own way. This is only human. However, I couldn't help but notice that nobody was thanking the surgical team. To be clear, dedicated medical professionals aren't in their profession to garner thanks. That's not why they took the job.

The danger I saw then, and that I see repeatedly today, is that, at least in the fundamentalist world in which I grew up, the scientific and medical community really played no role. It was, instead, in the most charitable view, God acting through the doctors. I see an ever-growing trend of outright science denial, of people now not only failing to recognize it was modern medicine who saved them, but taking it a step farther, and refusing to undergo treatment at all.

I've watched people of deep religious faith completely refuse life-saving medical care, put their faith in the Lord, and then pass away, leaving behind a mourning family and future grandchildren who would never get to know them.

How many more times in my life would I see supernatural

beings being given credit for incredible feats performed by medicine and science?

How many times would I see science and medicine shoulder all the blame when things went wrong?

Too many times. Far too many times.

Dinner with Friends

My early work in computer science and pharmaceutical programming, along with my pre-med studies, led me to contract work on an interesting research project with one of Kentucky's large universities. Although my dreams of becoming a doctor had succumbed to the realities of trying to balance a full-time career with the demands of a second round of college education, I was still very interested in medicine.

One of the truths of software engineering is that, if you are worth your salt, you cannot help but pick up knowledge from the subject matter experts you're working with. You can't write good medical software without being able to speak to the doctors and clinicians on the other side of the computer and understanding, at some level, what they're talking about.

Computer programs don't write themselves. Medical experts are rarely capable of producing quality computer programs. It takes someone who can walk, at least to some extent, in both worlds. And so, I was able to still participate in medicine to an extent, by proxy.

My manager on this particular research project eventually became a friend and, later still, a partner in a startup company with a third friend, an M.D. who specialized in pain care. I can see in my mind's eye readers of this book, more experienced in life than I was at this stage, shaking their heads and saying "Don't mix business and friendships." I was young then. I didn't know any better.

The work was exciting and provided opportunities that weren't available to most people. I was once again allowed to go on hospital rounds with doctors. Not for the first time, I was allowed to scrub in for surgeries—provided that I stayed in my assigned spot of course. I was an observer only, but I got to witness with my own eyes the miraculous opening of human bodies and implantation of medical devices that would alleviate suffering or keep someone's parent or spouse alive. These were heady days. I was "just" a programmer, but my thirst for knowledge was satiated.

Then one day it all came crashing down. My manager, my friend the doctor, innocently invited me to dinner with his wife. Not suspecting anything out of the ordinary, I immediately accepted.

It was a wonderful meal. My friend's wife, also a highly educated professional, happened to be an amazing cook. Dinner was delicious. We shared food, wine, and many personal stories. But then, my friend (and manager) cleared his throat.

"There's something we've been meaning to talk to you about. We'd like to share our personal testimony about Jesus Christ with you, if that would be OK."

This was totally unexpected and, had Diversity, Equity, and Inclusion (DEI) training been part of the workplace in the 1990s, would have most likely been a gross violation of our employer-employee relationship. And yet, we were also friends and business partners. As uncomfortable as I was, I saw no way out.

"The talk," as many of us who have since heard it would call it today, was predictable. I have no doubt it came from the heart and out of concern for my well-being. And yet it also came from a biased place—the speakers were assuming that their way of believing was the *only way*. Someone like myself who didn't agree with their train of thought was in peril.

I listened patiently to all of the arguments I'd already studied in depth for decades before this evening's dinner. As much as I appreciated friends reaching out in an effort to help, I really wasn't in any trouble. I could predict nearly everything they were going to say.

Except for one thing.

My highly educated friends, one with a doctorate in a specialized medical field, confessed that their marriage was based entirely on Biblical principles, including the belief that the man was the head of the household. Yes, they discussed things together as "equal" partners, but, in the end, when it came to final decisions, the very intelligent woman sitting before me always deferred to her husband. My jaw dropped as she said this.

I was stunned. I don't know why they decided to share this last piece of information with me. I had come to know and respect many religious people who rejected the misogynistic teachings of my evangelical church. They accepted science. They believed in God. They didn't preach to me. We had no conflicts. Now I was confronted by the person who held the keys to my continued employment trying not only to convert me, but putting me in the horrible position of necessarily criticizing his fundamentalist belief if I dared to speak my mind.

Having no response that would not sound dismissive and, more importantly, jeopardize my career, I bravely ran away. I had two choices. I'd accepted an innocent dinner invitation but was now put into a situation where I could either throw all of my religious and moral learning into a debate. Or I could simply thank my hosts for sharing their thoughts and thank them for a wonderful evening.

I chose the latter.

All of my educated, religious friends accept evolution. They accept that the Earth and the universe are billions of years old. They don't believe the ancient myth that all life on the planet was wiped out by a flood. They invariably reject teachings such as that LGBTQ people are somehow sinful. None accept the old Biblical teachings that men are superior to women. I respect these religious friends and family.

Quite often, they share the tragedy of having lost a loved one and understandably yearn for an afterlife where they can hold that person again; to tell them how much they are loved. This is a type of spirituality I can get behind, even if I'm not a believer. I thought that all my educated friends had thrown aside fundamentalist beliefs though.

At dinner on that night long ago, I learned I was wrong.

The Chiropractor

As strange as it may sound, I learned valuable life lessons from a couple of visits to a chiropractor. Being skeptical and science-minded, these were the kinds of "doctors" I'd normally avoid as if they carried the Black Plague. They weren't medical doctors. In the years I ran my science blog, *Bad Science Debunked*, I did a lot of takedowns of chiropractors and holistic healers who would claim every food under the sun was bad for you but then (you guessed it) turned around and sold their own products that contained the very same ingredients.

Decades of making a living as a computer scientist, sitting at a keyboard for forty to fifty hours a week, was taking its toll on my body. My arms and hands were numb. I was having trouble feeling my fingers. My neck was constantly in pain.

Friends who swore by chiropractors talked me into visiting one. I resisted for a long time, but was finally cornered when it was mentioned that I claimed to live by science. But here I was, rejecting something out of hand, without any evidence or firsthand experience. Trapped by my own words, I agreed to schedule a visit.

Things did not get off to a good start. I was laid out on a table where my legs were "measured." I don't mean via computer, lasers, X-ray, or even a simple tape measure. No, the chiropractor lined me up with some lines on the table, shook both legs to loosen them, stared for a minute, and then concluded that my left leg was shorter than the right.

I'd seen this trick before. It was very popular with traveling preachers, faith healers who'd lay an unsuspecting arthritic elderly person on a table, arrange their legs, and pronounce one as too short. You only had to see this once or twice before realizing that the man of God just wasn't lining up the poor person's feet. "In the name of Jesus, I pronounce you healed!" the preacher would cry. He'd tug one foot down to be even with the other and, amazingly, both legs were now the same length. Hallelujah!

Most likely from an adrenaline rush, the elderly person would hop down from the table and begin to dance while the joy band jammed and people clapped loudly and praised the Lord. Never mind the legs were the same length before the "healing." Never mind that once the emotion wore off, the person who was on the table was going to be in more pain. For the short while that the "patient" was on camera, the audience witnessed a miracle.

I just couldn't believe someone who claimed to be a doctor was trying to pull this poor magic trick off on *me*. "Don't worry," I was told, "your short leg is just due to a misalignment of your body. Given enough treatment, we'll get this straightened out and you'll be walking normally again."

But… I was walking normally when I came in. My problem was numbness in my arms and hands.

We weren't done. A wand that produced alternating current was placed on my arm, then moved along my skin as electrical signals were received and recorded on a computerized graph. The chiropractor claimed that the peaks and valleys we saw on the graph indicated problems in the underlying muscle and bone. Once again, I was promised, that after enough visits to the office, the graph would look entirely different and we'd see a much healthier result. Based on my ham radio and electronics experience, I had no doubt it *would* look different.

I knew that random "RF" (radio frequency energy) floated around me as I transmitted signals across the world from my bedroom in my ham radio days. RF is unavoidable. We get it from cell phone towers, television screens... even touching a computer keyboard or electrical appliance would cause a small, harmless current to flow randomly across one's skin. This wasn't news to me. For my science blog, I'd actually measured this current as part of a debunking of the expensive "grounding mats" and "grounding bed sheets" that became popular as part of the pseudoscience of "earthing." People were actually raking in $400 or more for metal-laced sheets that protected you from a totally harmless physical phenomenon.

"So," I asked the good doctor, "how do you know what's a healthy graph pattern and what's unhealthy?"

He thought for a moment. "Well, when you're healthy, the graph is going to look completely different. We'll compare before and after and you'll see a definite difference."

Of course, the graphs were going to be different. We were looking at completely random electronic signals. I wanted to get up and leave then and there, but I'd promised friends I'd see this through. And so, I stayed.

We went into the doctor's office for a consult on health and nutrition. This is the point where things were really going to go downhill, and I thought for a moment I'd be asked to leave the office.

"I notice you drink a lot of diet soft drinks. You know those are full of neurotoxins, don't you?"

"They are? Like what?"

"Like aspartame. It turns into formaldehyde. You know they

use that to preserve dead bodies, right? Do you want that poison floating around in your body?"

To some degree, he was correct. I knew from my pre-med biology and chemistry courses that aspartame broke down into methanol and the amino acid phenylalanine. That's why my soda cans carried a warning for phenylketonurics—people with a genetic anomaly that prevented their bodies from breaking down phenylalanine.

I didn't suffer from phenylketonuria though, so aspartame wasn't a worry for me. Much of the food the world eats—meat, fish, eggs—naturally contains phenylalanine. There's no difference between obtaining the molecule "naturally" through diet, or "artificially" via a soda.

But the formaldehyde? That was just a natural metabolite of methanol, and, of course, the human body had evolved long ago to remove small amounts naturally. Formaldehyde can be found in small amounts in several fruits, vegetables, and seafood. Doctors had discussed this in passing during my biology and physiology classes.

I calmly explained all this to the chiropractor and pointed to his list of foods that I should start eating to "cure" me. I quickly found what I was looking for: pears. Pears have trace amounts of formaldehyde. I asked if his formaldehyde was healthier than mine.

He became visibly angry. If I didn't want to take his advice, I could suffer the consequences. He was just there to advise me. He couldn't force me to live a healthy lifestyle. If I wanted to keep walking around on a short leg, so be it.

But... have I mentioned my visit was due to numbness in my limbs? I wasn't there for nutrition lessons. I wasn't a doctor, but had studied enough human biology and physiology to suspect that at some point we might actually get around to talking about my skeletal and nervous systems.

And, eventually, we did. I was taken to a room and asked to lay down on another small table. It was made up of a series of parallel wooden boards. He pulled a lever, and one of the small boards fell away underneath the curvature of my spine.

"Feel that?" he asked, one eyebrow raised knowingly.

"No," I honestly replied. "Didn't feel a thing."

I raised a knowing eyebrow at him in return.

"Don't worry. After enough visits, you'll begin to feel a difference. We'll have you walking straight in no time."

But I was walking straight when I came in! My office visit was because of numb…

Oh, never mind.

He had me sit up. He wrapped his arms around my head and neck and, in one of the most terrifying moments of my life, suddenly and rapidly jerked my head ninety degrees to one side. There was a loud POP.

"What the HELL was that?" I was too surprised and angry to watch my language.

"Just getting you properly adjusted and aligned. You should be good for today. I'd like to see you back two or three times a week while we're getting started. Please see the receptionist on the way out to schedule your next appointment."

I walked straight to my car. I was done with this.

And so I finally come to the moral of this chapter: my friends who swore by chiropractors told me the treatments obviously were never going to work for me because I didn't *believe* they would work. I can't count how many times I've been told the same thing about religion. I could only receive the knowledge imparted by the Holy Spirit if I *believed* first. God was no longer working in my life because I refused to believe. I wouldn't let him in.

Would God have saved my little sister from cancer if only I had believed? I don't think so. We studied various forms of cancer in school, and I don't recall a single instance where believing would have altered the inevitable outcome.

This way of thinking is contrary to everything I've ever learned in life. Science works whether I believe in it or not. To that end, I finally found a medical doctor and received a proper diagnosis. I was suffering from both cubital and carpal tunnel syndrome. The numbness in my hands and fingers were the result of a combination of pinched and compressed nerves in my elbows and wrists. Surgery and physical therapy eventually solved the problem.

Giving up diet soft drinks was never, in any imaginable world, going to help.

Golden Arches

The idea of living downtown in a large city was intriguing to me. I'd visited Europe enough to see that, with the right infrastructure, it was entirely possible. I got my chance when I landed a computer programming contract in downtown Louisville, Kentucky. There was a nice apartment building only half a city block from my workplace, a grocery store less than a mile's walk away, and plenty of restaurants and convenience marts for essentials like milk and eggs. I signed a lease and moved in.

Commuting to work was a breeze. The walk to my office took about five minutes. The only downside was I had no excuse for missing work on snow days. Then again, being one of the few people who could actually make the trip to the office, bad weather days were quiet and I got a lot done.

There were a fair number of homeless people in my neighborhood. I believe there was a Salvation Army or other type of shelter nearby. During the day, it was not uncommon to encounter people on the street asking for money to buy food. It was a personal policy to never just hand money over, but, believing in charity, if I had the time and money, I made it a practice to invite the petitioners out for a meal. I was surprised by the number of people who said no. Perhaps the money they were asking for wasn't for food—it was not for me to judge—but many others looked grateful and took me up on my offer. I met some very nice people.

I have no illusions that I was committing grand acts of charity in doing this. It just felt like the right thing to do, so I did it.

My apartment was adjacent to a McDonald's, so this was the restaurant at which we'd most often end up. Sometimes there'd be small groups of unfortunates sitting along the outside wall. Occasionally I'd see a familiar face and get a smile.

"Hungry?" I'd ask.

"You bet."

"I'll be right back. Extra ketchup, right?"

Sometimes restaurant management would get annoyed by the people hanging out and call the cops to run them off. They were bad for business, apparently.

Other times, I'd take people in with me. Once, a family of four. This seemed to anger customers and management. In one case, I had a worker come out and point at a man I'd brought in. "The manager says this is the last time he eats here. If you have any problems with that, you can speak to the boss. If you can't come in without him, then don't come in." He emphasized his point by pointing at the "we reserve the right to refuse service to anyone" sign.

I don't know what this particular man had done to be banned from a McDonald's. I didn't ask, though I did send an (unanswered) letter to corporate headquarters to complain. I didn't argue back. This is where I got my daily breakfast on the walk to work.

So be it, I'd just carry the food to people outside from now on.

Going outside presented problems of its own. At least once a week, a local evangelical church group descended on the sidewalk in front of the restaurant. They wore purple shirts and carried bullhorns and crosses. For hours (or at least until their batteries ran out), they'd harangue passers-by with Bible verses and warn them that they were in danger of burning in Hell. They sang hymns and danced on the sidewalk, while the hungry, homeless people looked on in something between amusement and bewilderment.

These fundamentalists were a regular fixture under the golden arches. They hung around for months. They were so loud I could hear their condemnations through the walls and closed windows of my apartment.

In all this time, never, not once, did I ever see a member of this church group walk over and offer to buy a homeless person a bite to eat.

Not a single time.

These people were here to preach and to condemn, not to feed the hungry.

Although by this time I knew in my heart I was an atheist, I still looked upon some of the teachings of Jesus with fondness. "Feed the hungry" was one of them.

Not to wax political, but I'd later recall my days of randomly feeding the poor: speaking purely hypothetically, what would happen if a throng of hungry, homeless people showed up on a river bank, looking for help? How would these purple-shirted evangelicals react? Would they hand out food and clothes? Or would they string barbed wire and quote scripture?

What would Jesus do?

Beginning of the End

My encounters with college evangelists like Brother Jed and Sister Cindy were but a small sample of the confrontations I'd have with fundamentalist theists after I stopped attending church. Friends, family, even total strangers constantly wanted to bring me back onto the "right path." The methods they employed inevitably had the opposite effect on me, driving me further away.

Personal testimonies had no meaning. I'd had chances by now to encounter a wide variety of people who'd made incredible claims that, I'm sorry, they couldn't offer evidence for—I'd have to just take their word for it. Sure, I'd had experiences that I couldn't initially come up with rational explanations for… until I sat down and thought things through. In the odd case where I couldn't explain something fully, I'd learned from many wise mentors to admit I didn't have an answer yet, and to keep searching for it.

Threatening me with Hell or withholding the promise of Heaven didn't get a well-intentioned missionary much further. Throughout my childhood and teenage years, I'd been tormented with threats of not only myself burning for eternity, but also the same fate for my loved ones who didn't accept my faith.

Honestly, it didn't even need to be a loved one who'd burn. I was brought up by loving parents, grandparents, and a supportive family who gave me a strong moral foundation. It eventually became impossible to reconcile any human burning in Hell while I enjoyed myself in Heaven. My morals would simply not allow me to participate in this belief system.

The very need for a creator to burn anyone, when that creator could have made any other choice, was a major stumbling block for me. The Bible taught me that we were all God's children. What parent would punish a child in such a way? I made my share of mistakes as a teenager, ignoring the advice and rules set forth by my parents, and arguably, sometimes I deserved punishment. But was there anything I could ever do that would cause them to torment me throughout eternity? Even if they'd warned me, and warned me, and warned me again?

No. Just no.

I would have the great fortune in my future to be a stepfather to the most wonderful daughter in the world. She could be a handful at times, and, having only met her when she was four years old, it took a little while for mutual love and respect to grow. But it did. It grew strong. To this day I refer to her simply as "my daughter." The word "step" seems to imply a lesser amount of love.

To speak plainly: There is nothing—nothing—in the world that my daughter could do that would leave me considering the type of punishment that God would eventually hand out to his children. It's simply unthinkable.

The crucifixion of Christ troubled me throughout my life, and his death became another "necessity" that I was forced to reject. As a little kid in Sunday school, I was horrified by the need for the Israelites to offer blood sacrifices. Why did our Heavenly Father require blood? To compound this problem, I was told that it was absolutely necessary for Jesus to stand in as a human sacrifice. Why? Why was a decent man forced to undergo one of the cruelest means of death possible, when all that was really necessary was for the person in charge, the one setting all the rules, to simply say, "I forgive you"?

The Catholic Church's teachings on original sin only made the problem worse. How could I, as a little baby, have been guilty of "original sin?" By the time I was deep into university studies, I'd matured enough as an adult to realize I was responsible for my own actions. I could not accept responsibility for sins created by someone who lived thousands of years before me, any more than I could assume the guilt for people around me in school.

I could no longer follow in the path of my fundamentalist upbringings. I could find no comfort in other denominations, such as the Catholic faith I was so familiar with. More moderate Protestant denominations offered no answers to the conflicts I was experiencing.

The realization didn't hit me in a single moment. It had been creeping up on me throughout the years. My garden of doubt, the first seeds of which were planted because of a theological threat over a simple prism decades before, was ready to harvest.

While I was not going to press my decision or doubts on any other person, I could no longer call myself a Christian.

And then... the End

As I entered my thirties, life began to change in subtle ways. The imagined immortality that seems to infect everyone aged twenty years and younger began to fade away. I started losing family members, including my beloved Mamaw, who was like a third parent to me. A close aunt, who I'd lived next door to my entire life, soon followed down the path we must all someday travel. I began to think about death. I had never really completely stopped religious studies in my life—we had a great local library in Sunnyvale, CA, where I passed my days in Silicon Valley mostly free of religious indoctrination. I'd still visit and do some reading.

But now thoughts like death, eternity, and a possible afterlife occupied my mind. It's almost like I woke up one morning and realized I wasn't going to live forever. That 120 MPH sprint I did on a lonely Nevada highway in my Firebird, back when I was in my early twenties? Yeah, that kind of thing was out now.

I still felt that there must be something—something I couldn't explain—tying the whole universe together. Though no longer a believer, I do hear this sentiment often from religious friends, and I understand the feeling. I may no longer hold the belief, but I can appreciate from where it originates.

At this point I was certain I was no longer a Christian. I was finally able to admit this to myself. Though I had many people in my life who followed what I agreed were good teachings of Jesus (taking care of the sick, the hungry, the poor), I could not bring myself to believe that someone needed to die a horrible, bloody death in order for me to be forgiven for wrongs that I myself had committed.

The scars of fundamentalism, all of the science denial, left marks on me that would not go away. I would take the good things I'd learned in my religious upbringing, discard the bad (e.g., a 6,000-year-old Earth), and I would move forward.

Of all the other religions I'd looked at, Judaism offered some possibilities. There was no Heaven or Hell to worry about. I could still believe in a God, but I could question him. And I did have questions. There were a lot of evil things that went on in the Old Testament, and I couldn't excuse them. So I would investigate. I would research. I would see if there were answers that made sense.

I enrolled in conversion classes, at both Reform and Conservative synagogues in Lexington.

Judaism classes were enjoyable. This was one religion that had never reached out to me or my classmates and invited us to convert. We were there of our own free will. An enthusiastic class with dedicated teachers is an exciting place to be, regardless of the subject matter.

I was especially fascinated by learning to read Hebrew. This was the foundational language of my very first religion. Nobody I'd ever gone to church with as a youth could read Hebrew. I was growing my childhood knowledge.

There was no difficulty in learning to read right-to-left, as I had feared. I'd always had a decent memory, so picking up a new alef-bet (this is how we write "alphabet" in Hebrew) came relatively quickly. It was the fact that there were technically no vowels that gave me pause. Yes, for we beginners, vowel marks were added, but if I was ever to graduate to "real" Hebrew, I'd just have to know where the vowels were supposed to go.

It was a joy reading Torah verses in our Hebrew primers that I knew in English by heart from childhood. But when I went out and Googled images of the Dead Sea Scrolls (vowels not included), I knew I had a lot of work ahead of me. Just because I could read and pronounce Hebrew words, it didn't mean I knew what they meant.

Both the reform and conservative synagogues I visited were open, inviting places. People invariably noticed the newcomer and welcomed me warmly. And yet, I couldn't help but feeling somewhat like a fish out of water. I really didn't have a history to fall back on. It was probably shyness more than anything else, but I didn't have the sense of community I remembered from my youth. I was realistic. I didn't expect such feelings to come overnight. It would take time. I continued to attend services and classes and tried to absorb all I could.

The customs and rituals, which I could tell bound a community together throughout time, didn't hold a lot of meaning for me. I had already been shedding such accoutrements throughout my journey away from Christianity. A mezuzah on my front door? Sure, I could do that, but honestly it was never going to have the significance it would to someone who'd grown up with the symbolism.

Though I've forgotten much of my Hebrew by now, I reached the point where I could read along well enough to pronounce and sing along with songs. In our conservative classes, this was a favorite part of the rabbi's class. I was the only person in a very musical family not blessed with a good singing voice, but I tried my best.

I *did* feel something as I sang along, or read ancient prayers like the Shema, but I don't know if the experience could be described as spiritual. It was more the wonder that I was reading and singing the same language that others had used thousands of years before. A lover of language and history, My excitement probably came more from intellectual stimulation than religious.

As I sat with the congregation during Shabbat services, I felt welcomed by the group, but I didn't feel spiritual. The companionship I gave up when I left my fundamentalist church was back, and it was important. But I didn't feel the connection to a higher power that I so desperately sought.

Words from my experience with the chiropractor came back to me… perhaps my failure to believe was because I didn't already believe. Being a born skeptic, tautologies and I never got along well together.

After a long period of *trying* my best to feel something, *wanting* to be part of a religion, and *desiring* to believe, it dawned on me that I was repeating the very same mistakes of my childhood. I was trying to self-inflict an emotional response that just wasn't there. The difference in this case was that I wasn't surrounded by others urging me on, promising me eternal life or threatening me with eternal punishment. I did love that about Judaism. The lack of proselytizing was refreshing. I was able to make a decision based on my own knowledge; my own life experiences. And so I did.

I had to be honest with myself.

I just didn't believe in gods anymore.

This was not an easy admission to make to myself. The fundamentalist teachings I grew up with still haunted me, even after I had intellectually thrown them off. I've since read many accounts of survivors of indoctrination and find I have much in common with them.

There was guilt—I'd let my religious family and church down.

There was the lingering fear of punishment—I'd spent over a decade crying myself to sleep over fear that I or someone I loved was going to Hell.

There was loneliness and isolation—I was a minority, a small non-believing minnow in an ocean of fish that all believed. Some of those fish were bigger than me, figuratively speaking. They held the keys to future jobs, memberships in social organizations.

There would be myriad reminders that I was different. Somehow, people would notice that I didn't recite a certain pledge to which the words "under God" had been added in the 1950s. I didn't bow my head for prayers... people would always see this (paradoxically so, because they should have had their heads bowed!).

I despised the word "atheist." I was trained very early on that this was the absolute worst thing that anyone could be. The word only literally means "one who lacks a belief in gods," but it came with years of fundamentalist baggage. I avoided it at all costs. Initially, I kept quiet about my decision.

I knew from my own ostracism of non-believers when I was a fundamentalist, that the world was going to be a very different place for me from then on.

Isaiah 7:14

Even though I came away as a non-believer, my time in Judaism classes was not wasted. In addition to gaining a better understanding of another major world religion, I found myself able to stumble over basic Hebrew. There was a certain... power? in being able to read words as they'd been originally written thousands of years before; words that had played such a powerful part in my upbringing. I didn't need to be religious to appreciate reading words written by the ancients... words that, for better or worse, would affect history.

Foreign languages have always fascinated me. I got an introduction to other languages as a ham radio operator in my teens, muddling through conversations with people in South and Central America with broken Spanish I'd learned through a mail-order course.

I moved on to French in high school, and became proficient enough to converse with new friends in France—even doing it through Morse Code just for the added challenge. Actually, being a Southerner, my French accent was so horrific that it's just as well we were conversing in code. The French are justifiably proud of their language and, on later trips to French-speaking countries, I'm sure that I caused quite a few chuckles.

I continued French studies in college and by the time I graduated I was fluent in the language. It wasn't until later in life, when I had forgotten a lot of what I'd learned, that I got the chance to actually use the language in France, Niger, Montreal, and French Polynesia. Hats off to all the native speakers who tolerated me—I could see the puzzled grins on their faces as I blurted out what I thought was decent French, but they seemed flattered that an *American* was actually making the attempt. Here was someone from the USA who didn't labor under the misapprehension that shouting the words more loudly in English was going to make them understandable.

I loved Hebrew. Once I got the alef-bet down, it was as easy as reading English—except that I had absolutely no idea what I was saying. Sure, I could pronounce "מֶלֶךְ" ("Melech"), but I had no idea it meant "king." So while I could claim after a month's classes that I could read Hebrew, it was only *technically* true. I could sound out the words; I just needed to learn what all they meant.

This is leading to a point, I promise.

One of the ways the rabbi leading our conversion classes made things interesting for us is to pick verses from the Old Testament that the class, most of whom were or had been Christians, would be familiar with. It was a thrill reading part of the 23rd Psalm in the original Hebrew. I felt a connection with my past church life. For all the trouble it brought me, there were times when I found comfort in religion. Despite the distress I've described, I did find short-lived moments of peace.

In Hebrew class, as we broke down the sentence structures and what the words actually meant, my French and Spanish education helped. A really common misconception among Americans in the "speak English damn it!" crowd is that the words in a sentence in one language are all just piece-by-piece replacements for words in the same sentence in another language.

It doesn't work that way. With apologies to any linguists reading this book for oversimplifying, phrases in a given language are an arrangement of words that, when put together, have some particular meaning to the listener, who understands that language.

To convey an English thought in Hebrew, or vice-versa, a translator has to consider the thought to be conveyed and come up with the best words in the target language that would convey the same idea.

And so I finally come to the point of this chapter. One night in Hebrew class, we were reading one of the most famous verses in the Old Testament: Isaiah 7:14, the verse that prophesied the virgin birth of Jesus:

"Therefore the Lord himself shall give you a sign; Behold, a virgin shall conceive, and bear a son, and shall call his name Immanuel." (Isaiah 7:14, KJV)

This was one of the foundational verses of my religion. Despite being in the Old Testament, it was often trotted out at Christmas to validate the circumstances of Christ's birth.

Of course, we weren't presented with the words I was familiar with. What we saw was:

לָכֵן יִתֵּן אֲדֹנָי הוּא לָכֶם אוֹת הִנֵּה הָעַלְמָה הָרָה וְיֹלֶדֶת בֵּן וְקָרֵאת שְׁמוֹ עִמָּנוּ אֵל

As we worked our way through this verse word-by-word, alarm bells went off in my head. It was all caused by one word: הָעַלְמָה ("ha-almah").

This word means "young woman," not "virgin."

Our rabbi explained that this was one of the most controversial verses in the Bible. And it was all due to scribes substituting the wrong Greek word when translating the Old Testament in or around 300 B.C. I was on familiar ground here. I once made a translation error in Prague that nearly got me arrested. Yet another story for another time.

But I digress. As I sat there reading the original text of Isaiah, it was obvious to me that I was looking at a translation error. This was certainly a discrepancy that would have never been brought up in my old church. I would have been condemned for mentioning it.

As I learned in religious studies at university, a great deal of the religious texts available to early Christians were written in Greek, the predominant language in their part of the world during the nascent years of Christianity. An important translation, known as the Septuagint, was the primary source of Hebrew-to-Greek. And, unfortunately, some of the early Septuagints mistranslated the Hebrew "almah" (young woman) to the Greek "parthenos" (virgin).

I say "some" of the Septuagints because, unknown to some Christians (and certainly unknown to me as I grew up in the church), there are multiple copies of this document, and not all of them agree with each other. Some Greek translations don't use the word "virgin." They correctly read "young woman." But the version floating around the Holy Land in the times leading up to Christianity must have had the mistranslation, and it set the stage for what became so obvious to me as I sat there that night in Hebrew class.

Early Christians, with only the error to go by, would have expected a virgin birth. As believers, perhaps they would have written this into the New Testament texts they'd come to produce.

In the back of my mind, I could feel my fundamentalist church condemning me for having these thoughts. However, just as they refused to study evolution, so had they apparently decided not to study the original language of their scripture. They took it all on faith.

The correct Hebrew word for virgin, by the way, is בתולה

("betulah"). It's actually used in the same book, Isaiah, as the mistranslated verse 7:14. In context, "betulah" is used to describe a man's joy over his marriage to a virgin bride. See Isaiah 62:5.

When I was a young child enduring my early years in the church, I was given the impression by my teachers that the Bible was one complete book, divinely inspired by God, meant to fit together perfectly, and was error-free. Translation mistakes were impossible according to my fundamentalist teachings.

I must pause here and reiterate to the reader that my goal is not to attack a particular religion or a belief. Honestly, it would not have mattered to me at all if Jesus was of virgin birth. I can still accept the incredible beauty and meaning of the words he said in his Sermon on the Mount. The problem I'd like to shed light on is what happens when fundamentalists tell you a translation is perfect... and then you find out it isn't.

This is one of the fundamental (no pun intended) problems with fundamentalist organizations such as Answers in Genesis. They paint themselves into theological corners by insisting on Biblical inerrancy. When you're told, under threat of eternal punishment, that every word of a book is literally true; that not one single sentence can be incorrect, or the whole thing is incorrect... well, the problems are obvious when you find an error.

As life went on and I became somewhat literate in other languages and the nuances of translating them, it did seem to me that some interpreters of more modern versions were guiding the process to make the Bible seem more "true." I've mentioned that part of my early education was reading and re-reading the Bible cover to cover. In doing this, I discovered and read versions more modern than the familiar King James, and made fascinating discoveries.

It was my interest in astronomy that clued me in to what seemed to be an attempt to make modern translations more "in the know." Look at Job 38:31-32, KJV:

31 Canst thou bind the sweet influences of Pleiades, or loose the bands of Orion?

32 Canst thou bring forth Mazzaroth in his season? or canst thou guide Arcturus with his sons?

How do you bind "sweet influences" of the star cluster named the Pleiades? That's as unscientific as you can get.

And as for guiding "Arcturus and his sons?" Apparently, a reference to other stars near Arcturus that seemed to follow it across the night sky.

I went to a more modern translation, the New King James version, written in 1982:

31 "Can you bind the cluster of the Pleiades, or loose the belt of Orion?

32 Can you bring out Mazzaroth in its season? Or can you guide the Great Bear with its cubs?

Can you spot the differences? Notice that the Pleiades are now correctly referred to by the modern designation of a "cluster." This is nowhere in the original Hebrew. More telling is that Arcturus, which wasn't even a star known by any name to the Old Testament authors, has taken on its Western nickname, a reference to "Guardian of the bear."

And what happened to the "sons of Arcturus?" The original Hebrew does refer to sons/children, but the 1982 translation leaps on Arcturus being associated with bears and replaces "sons" with "cubs." Cubs would naturally follow a bear, right?

But the Hebrews who wrote this verse didn't have a "bear star" and never said a word about cubs. Modern scholars don't even know which star Job is actually talking about. Although Orion and the Pleiades were well known to the ancient Israelites, they never said a word about Arcturus. My conclusion: modern translators were taking liberties and "enhancing" verses. Perhaps this was done with the innocent intent of making ancient texts more approachable, but in talking to fundamentalist evangelicals, I noticed something completely different happening:

"Oh look! Job knew so much about astronomy that he correctly identified star clusters and predicted little cub stars following a mama bear, before astronomers even called it that!"

Yes, I've actually had this discussion with fundamentalist friends, and they do indeed credit Job with far more knowledge than the original Hebrew grants him.

I began this chapter with an examination of a significant translation problem for people who believe in prophecies and virgin births. I ended with a more light-hearted example of how translation can be a tricky business, and, if one is not careful, lead the reader to conclusions that just aren't there in the original text.

Of the two verses we've looked at, Isaiah 7:14 was a show stopper for me. By the time I started my (admittedly very simple) studies of Hebrew, the Dead Sea Scrolls had been discovered and the entire original text, The Great Scroll of Isaiah, was known. You can find images of the thousands-of-years-old text online and, with a little effort, read them for yourself. My intent is not to sway someone from a belief; rather, it is to challenge the fundamentalist perspective that every single word of scripture is inerrant.

The Septuagint, being a translation of another language, means the Greek translation came after the original Hebrew. You can't have a basic understanding of language and not see the error here. You just can't. Certainly, this wasn't intentional. The Septuagint was written around 300 years before the birth of Christ. In no way do I mean to imply that anyone was "stacking the deck" for a virgin birth. It looks like an honest mistake. Translation is not an easy job. Nobody is perfect.

Going back to my earlier chapter titled "Baptisma," I mentioned that my fundamentalist Sunday school teacher, with no knowledge of Greek, insisted that the word "baptisma" meant total immersion in water. I'd later pick up a few words of the language and learn that it simply meant "cleaning" or "cleansing." But this one word was enough for my evangelical church to condemn both my mother and I for our Catholic baptism, which involved just a sprinkling of water.

I thought back to countless nights of terror, trying to fall asleep, while worrying my mother might not make it to Heaven. All that fear because someone didn't translate correctly.

Words matter.

Through basic studies of languages, I learned that scripture was not as black and white as I'd been told.

9/11

Every American remembers where they were on this day; what they were doing. It was a warm Fall morning in Kentucky. My father and I were loading our camper for a week-long photography trip to the Smoky Mountains. The phone rang. Mom was calling from work. She sounded very panicked.

"Turn the TV on. Something's wrong. They're saying something about a plane crash."

I turned on CNN and was greeted with a live photo of what I immediately recognized as one of the World Trade Center towers. Black smoke was pouring from one side. Flames were visible along the edges of what looked like a huge gash in the building.

As I watched, the camera cut away to another angle and I watched in horror as a huge jet crashed into the other tower. A huge firewall, a wave of flames, rolled through the building.

Reporters were speechless. Nobody knew what was going on. Reports were coming in of planes hitting other buildings. I felt panic. I felt numb. I didn't know what to feel. Dad, wondering why I'd stopped helping loading the camper, walked into the house, saw the screen, and sat down. We didn't say anything.

It would take a while before the story came out, but of course it's the one we all know so well now. Religious fundamentalists had flown commercial jets into buildings. Their God had told them it was a just act. Live shots from cities in the Middle East showed people dancing, singing, praising God, handing out candy.

What was happening to the world?

There was a lot of anger toward Islam at this time. I was angry too but tried not to paint with too broad a brush. I'd seen fundamentalism from many religions by this point. I'd been part of a fundamentalist church. It wasn't that hard for me to imagine how young men, indoctrinated and radicalized from birth, could be convinced to carry out such acts. I was more focused on how fundamentalism in general could lead humans to act in a certain way, rather than narrowing in on a particular religion.

I recalled the lesson of Abraham and Isaac from my childhood... the lesson that someone could believe God gave the okay to kill someone, and the person who would do the killing was considered good and moral. The takeaway for me was that someone could be so blindly obedient to their creator that they would sacrifice other people on command. In a context my evangelical church would agree upon, this was a righteous thing.

On 9/11, I heard echoes of my own religious education. When God tells us to kill, we obey.

Street Corner Debates

As it became apparent to me that I was progressing more deeply into the realm of not believing in gods or the supernatural, I was initially introverted when it came to religious discussions. I would avoid them. I was painfully aware that I was a minority and, according to some Pew research, a member of the most mistrusted group in the United States.

Yet the constant, unwanted pressure from outsiders to repent, convert, to see things in their own way, grew more and more oppressive. There was nowhere I could go and avoid all the noise. Family reunions were full of well-meaning relations, who wanted to talk to me about why I no longer attended church. Knowing the ostracism inflicted on outsiders from seeing how my own religion treated "outsiders," I feared what awaited me if I came out with the truth. And so I demurred, telling people "I just don't believe what you do, and I find it better not to talk about this subject."

I finally reached a breaking point though. Living and working smack in the middle of downtown Louisville for nearly a decade, I ran into not only the homeless, who I regularly fed, but also streetcorner missionaries, who were looking to save my soul rather than get a bite to eat. One day, when a friendly face approached me and recited a Bible verse, something snapped— and I recited a verse back at him. A verse from Psalms about killing children. His smile dropped. He wasn't familiar with the verse. In for a penny, in for a pound, I invited him to look it up. And he did.

"Oh, see, you're taking this out of context," he replied. And so

I gave him the context. The first signs of doubt appeared on his face. A debate ensued.

Not to paint with too broad a brush, but in this "debate "and countless others that have followed, I discovered that invariably, the person trying to do the converting isn't prepared to run up against someone who is intimately familiar with the subject matter. We went at it for fifteen, maybe even thirty minutes. I remained calm and pointed out errors in his line of thinking, sticking with scripture; we often had to pause so he could flip through his Bible and look things up. He seemed to grow more and more frustrated, perhaps even a bit angry. It finally dawned on him this was going nowhere.

"Brother, I am going to pray for you. I hope you'll see the light." He gave up and moved toward another passerby.

I am the first to admit I love a good debate, though before this occasion, I never sought it out when the topic was religion. I avoided theological arguments like the plague. Any other topic— science, politics, you name it—I absolutely loved diving in. It wasn't from the arrogant or narcissistic perspective that some might falsely assume. Throughout my education, I was taught that debate was a legitimate way to advance my knowledge. Quite often, I'd find myself having to (gasp!) admit I was wrong and go back and re-evaluate a position.

But, regarding religion: Why, for all these years in my life, should I have been holding back my thoughts on a topic that everyone else around me felt free to bring up at the drop of a hat? People would literally chase me down a street to preach or pound on my door in the middle of a critical NFL playoff game just to get me to listen to how I could be "saved?"

Why was it that everyone in the world was free to make daily Facebook posts exclaiming the wonders of their faith, but I was forced to remain quiet so that I didn't offend anyone?

Why did people in our society feel free to ignore what I was taught was proper decorum ("don't discuss religion!")? As I mentioned in the chapter "Dinner with Friends," this would happen to me more than once, even in professional settings. Friends, family, even my employers, felt absolutely no qualms over telling me all about their beliefs, and why I should follow them.

During my computer contract stint in Louisville, I shared a cubicle with a very likable young man who happened to be a Jehovah's Witness. We got along very well and often had lunch together. He had a fascinating scientific mind. One of his projects was to make weekly rounds to all the restaurants on Fourth Street (Louisville's main entertainment drag) and collect all the used cooking grease. He'd dutifully cart it home and convert it, in his garage lab, to fuel he could burn in his car. Until the health department prohibited restaurants from giving him the raw materials, based on grounds they weren't following city code and disposing of waste properly, my "JW" friend paid very little for gasoline. He also drove a car that smelled a lot like French fries.

But he was also passionate about his religious beliefs, the discussion of which were very clearly prohibited by our employer within the confines of work.

And yet this friend still found a way. Quite often, far more times than could be deemed a coincidence or an "accident," he'd leave behind religious tracts in our cubicle. Sometimes I'd find a copy of the Watchtower on my desk. "Oh, I'm sorry, I forgot where I put that," was his excuse. I'd get spontaneous invitations to "exciting life-changing seminars." No, thank you, but I just don't want to go to church with you. I meant no malice; I just wanted to be left alone.

I finally had to have a frank discussion with my buddy in our company cafeteria. He seemed genuinely shocked that I was a total non-believer.

Our friendship cooled after this. I knew from other Jehovah's Witnesses who weren't quite as adherent to their religion that they were supposed to avoid all contact with what my Muslim friends would refer to as infidels. If I wasn't a plausible target for conversion, then I was to be shunned.

And yet, after all of this, I'd still occasionally find a copy of the Jehovah's Witness publication, "Watchtower," "mistakenly" left behind in our shared cubicle.

Becoming more outspoken about my lack of belief, I discovered that my open rebuttals and confrontations with those who approached me without invitation were bearing fruit. Once people knew my position, and found they couldn't "out-Bible" me (all those decades of scripture study paid off in an unexpected way), conversion attempts decreased.

Unfortunately, my social circle also shrank. People did not want to associate with a non-believer. I did understand this—it's exactly what I was taught to do over the years as I was indoctrinated as a youth. Many fundamentalist religions (and cults) strongly discourage fellowship with those who don't share their faith. The reason for these prohibitions became clear to me:

When you're free from contrary opinions and evidence, and spend time only with those who think as you do, it's easier to keep the flock together.

The Wild, Wild Web

Armed with my newfound confidence in taking on those who'd engage with me on science and theology, I began seeking out opportunities to exercise my freedom. I had the same right to speak my mind on a subject I'd once considered socially taboo, as did those who were preaching to me. Under no circumstances did I ever reach out and try to "convert" someone to atheism. I had many wonderful friends and family members who were religious. They didn't feel the need to press their beliefs on me. Likewise, there was no motivation to try and sway them toward my non-belief.

But the blatant proselytizing, the religious attacks on science, even the cherry-picked Bible verses from books such as Psalms, where I knew dreams of blood vengeance were separated from "The Lord is my Shepherd" by only a handful of pages, became fair game. I was happy to stay in my lane of science and logic. When someone crossed over into my lane though, I felt it proper to sound my horn.

In retrospect, perhaps one of the biggest mistakes I made was speaking out on social media. I was weary, oh so weary, of (literally) daily posts from hundreds of friends proclaiming that everyone else should share their faith.

I recoiled from Facebook posts proclaiming miracles when a single person survived a tornado wiping out all but one person in a home, or a single person survived a bomb attack, or an airplane crash in which hundreds of others died. All but one person died in these tragedies. How was this a miracle? In the eyes of many, I became "the angry atheist"—someone who would actually post a reply and point out that these were not miracles at all. Was I wrong to inject logic into unsolicited posts that appeared in my social media feeds? Was my opinion less valid? I wasn't trying to convert anyone. I was pleading for basic logic in a world that seemed to be slipping away from reason.

In truth, I did harbor anger over my religious upbringing. At my advanced age of sixty, I have no qualms in pointing back to the indoctrination children underwent (and are still undergoing) in fundamentalist evangelical churches, as abuse. Perhaps the tone of some of my comments could have been less confrontational.

I definitely did not attack or respond to every religious post I encountered. I felt warmth from the posts by family and friends from their places of worship, celebrating the joy and spiritual fulfillment they get from the experience. I understand and rejoice with those who give thanks for the simple blessings of being together with family, welcoming a newborn into the world, and so on. I'm happy when I see others happy.

But something in me was changing. I felt, more than ever, the need to speak out. Even in the 1990s, I could see society slipping toward the fundamentalist, right-wing attitudes that are front and center in so many disagreements today in 2024. One of my favorite quotes, attributed to Christopher Morley:

"Seek argument for argument's sake. There will be plenty of time for silence in the grave."

The 1990s and 2000s provided fertile ground for honoring Morley and seeking out argument for argument's sake. The World Wide Web and social media had consumed the world. There were online "debate" forums for every topic imaginable, including religion, and, the special area of interest to me, the conflicts between science and theology. I put the word "debate" in quotes because, as anyone who's ever engaged in these online arguments will admit, there's really not a lot of formal debating going on. I can't recall a single instance of anyone changing anyone's mind.

So why did I engage in this behavior? For the same reason I read Grandma's National Enquirer and other tabloids in my youth: to sharpen my own knowledge and critical thinking. It's easier to argue against a position if you know what that position is. As a youth I learned to apply critical thinking to the mystical, miraculous claims and prophecies found in tabloids such as The Globe. I now found myself free to apply this same thought process to Young Earth Creationist claims. I discovered fertile ground for my skeptical musings:

A new enemy had arisen in my own figurative backyard of Kentucky. A dark and dangerous enemy. A group named "Answers in Genesis (AIG)."

AIG had erected, with help from the taxpayers of the Bluegrass State, two so-called "museums" that can only be fairly described as shrines to pseudoscientific thinking: the Creation Museum and the Ark Encounter. I engaged AIG followers early and often in online debate forums. As in real life, so it would be in the virtual, online world. I never expected to change the mind of a single believer I argued with. Many scientists would say that it was perhaps folly to even interact with such people at all. Thomas Payne perhaps summed it up best:

"To argue with a person who has renounced the use of reason is like administering medicine to the dead."

But I found value in such debates, if they could be called that. American society seemed to be inexorably sliding toward a state of constant science denial, while fundamentalist religious beliefs were gaining an alarming amount of ground among certain politicians, and finding footholds in law; being taught in public schools. The beliefs being pushed here were the very same ones forced on me, via threat of eternal suffering if I denied them, as a youth.

With the birth of the Wild, Wild (World Wide) Web, Young Earth Creationist thoughts could be spread across the planet at the speed of light. If I was going to speak out against these teachings, I needed to recall the lessons I learned as a child in church. I had to understand how the teachings had evolved, and how the secular science I'd studied in the intervening years was now being horribly twisted by creationist organizations.

I realized that not only was I wrong in not speaking out during all those years I'd avoided debate with religious fundamentalism; not only was I correct in breaking my silence and openly engaging those who sought to deter those who wished to walk a path of reason…

I wasn't speaking loud enough.

Ark Encounter

I've always enjoyed science-based writing and have been fortunate enough to have had my work picked up by publications and organizations that I enjoy and admire. One such entity is the National Center for Science Education (NCSE) who, perhaps through my work with skeptical magazines or my blog, *Bad Science Debunked*, contacted me and asked if I'd be interested in writing an article for them.

At the time, the Answers in Genesis organization was touting its brand-new Kentucky theme park, The Ark Encounter—a life-size recreation of Noah's Ark. Science-minded Kentuckians were already embarrassed by an earlier AIG construction, a Creation Museum, which falsely claimed that the Earth was only 6,000 years old and humans lived alongside dinosaurs.

NCSE was covering the attraction in an upcoming issue of their newsletter and, being a former Young Earth Creationist who had found reason in science, they wondered if I'd be interested in writing a perspective from the viewpoint of someone who'd been there, done that, and bought the T-shirt. I said yes immediately.

This did mean actually visiting the Ark Encounter, something I was not entirely comfortable in doing. First, I didn't want my money to go toward supporting an anti-science cause. Second, I was worried about triggering memories that had left indelible scars on my psyche. However, I saw it as absolutely necessary. I often criticize anti-evolution forces who have never actually studied science or evolution, so it would be hypocritical of me to write about a place I'd never visited—even if I knew the Biblical background well.

And so it was that, on one sunny spring day, I found myself pulling into a half-full parking lot that faced a large wooden boat that, to my eye, would have been falling apart if not for all the giant support structures holding it together. I placed my "Ark Parking" ticket on the dashboard of my car as instructed, and got in line for a bus to the vessel. But not before passing through a conveniently placed gift shop.

My tour began by standing with a crowd and watching a short video concerning the great flood's backstory. I was already in full skeptic mode. I wanted to ask our tour guide why God, being all-powerful, couldn't have simply had all the people and animals just drop dead. He'd already wished them into existence once. Why not just blink and start all over again?

I wanted to emulate William Shatner's James T. Kirk in the movie, *Star Trek V: The Final Frontier*. Confronted with an evil alien entity posing as God, Kirk has a prescient question when "God" asks him to bring the USS Enterprise closer so that he can make use of it:

"Excuse me... Excuse me... I just wanted to ask a question. What does God need with a starship?"

I had the same question on the tip of my tongue: "Excuse me. What does God need with a large boat?"

Or why did our creator see fit to drown children? Why go to all the trouble of flooding the whole planet? I held my tongue. I had an article to write, and I didn't want to get kicked off the property before I'd even boarded the boat.

The first feature I noticed as I came aboard took me right back to the most frightening parts of my early childhood. Satan, in the guise of a large red serpent, looked directly into the eyes of visitors and informed them that if he could convince you that the flood wasn't real, he could convince you that Heaven and Hell weren't real. How many times had I heard this in church and Sunday school, and felt afraid to question my instructors? I looked around. There were a lot of young children in line with me. My heart went out to them. "I too was once like you."

The message was clear. Ken Ham and the Ark Encounter were following in the footsteps of my fundamentalist church, wrapping themselves in the cloak of the word of God, making it impossible to deny anything else you saw or heard.

The indoctrination would begin on the ark's gangplank, and it would get worse from there.

Honestly, the exhibits were underwhelming. There were a lot of automated dioramas where Biblical characters tried to act out stories but came across as malfunctioning robots from the movie/TV series *WestWorld*. I half expected one of the character's faces to pop off like Yul Brynner's did in the original movie. The skeptical part of me actually prayed it would happen.

My prayer went unanswered, and I moved on.

The ark "science" has been so thoroughly debunked by so many writers that I'd be beating one of Noah's two dead horses if I tried to cover it all again here. I'll just touch on a few items that either hit me as funny, contradictory, or disturbing.

One odd display involved polar bears. "How did polar bears survive when they live in cold climates?" was the science up for debate. Well, the plaque informed us, it's a misconception that polar bears need a cold environment. It had pictures of polar bears frolicking in the California sun at the San Diego Zoo. "Ah! So that's how they did it!" exclaimed a man to his son, both of whom had walked up behind me. "I'd always wondered about that."

Should I stop him and tell him that the San Diego Zoo has a carefully monitored air-conditioned environment for their polar bears, and they are treated to a chilled plunge pool? Polar bears are most definitely not suited for warm environments. No. Just bite your tongue, Mark, and keep walking. The man stepped in front of me and pointed at the bears frolicking in the plunge pool.

"Doesn't this explain so much?"

Well, it did if you knew the pool was artificially cooled. But I wasn't here for debate. I pasted on a fake smile and started to walk away.

But then I paused. Another thought hit me: this entire exhibit was a complete and total violation of one of AIG's core arguments. As Ken Ham would tell you, Noah didn't need to take two of every animal. He only needed to take two "kinds" of animals. Now, a "kind" is as far from a scientific definition as the North Pole is from Mount Ararat, but, regardless, bears qualify as a "kind" and, according to AIG, all that was needed for the ark were *any* two bears.

There would have been no need to take polar bears. According to AIG, the two "bear kinds" would have evolved into polar bears, grizzlies, pandas, and so on.

Oops. Evolution. Yes, AIG admits it happens, but only on time scales and in circumstances where it suits their needs. At all other times, evolution is a "religion."

I escaped the bear-loving man and ran straight into a cage containing two dinosaurs. I couldn't believe it. Yes, I know that Ken Ham claimed that these creatures were on the ship, but it still seemed like a joke to me. But there they were—baby dinos. Well, plastic molded models. Apparently, the animatronics department at AIG had a budget shortfall, so the dinosaurs couldn't move. They just sat there in their crates, looking artificial and out of place, just as they would have been on a real ark.

We walked through what was apparently the food storage section. It looked incredibly tiny. I couldn't convince myself there was possibly enough room to store enough food for all those animals for an entire year. Later, long after my NCSE article was completed, I was able to sit down with an AI engine and do some back-of-the-envelope calculations on how much storage would be required for varying amounts of "kinds" of just the herbivores on board. Based on Answers in Genesis' own estimates of 6,700 sheep-sized animals on this vessel, I couldn't get the number down to less than 100% of the entire ship even under the best of conditions, and the range was as high as 300% of the vessel in the upper case.

One mistake that the Ark Encounter creators made is that they claimed Noah took only baby animals to cut down on space requirements. For mammals, this isn't possible—they haven't been weaned yet. Without mothers to provide milk, the babies would die. I looked at the baby animals (again, sadly, no animatronics), shook my head, and asked, "Where's your mommy?" Oops. I was talking out loud. People turned and stared. A bright young girl walked over to me and explained that there were no mommies—there was room for just two of each kind. Dare I stop and explain about weaning? No. Just keep moving.

I ran across another gift shop. This place seemed to have more such shops than Disney World. The stuffed baby dinosaurs were selling like hotcakes.

Of course, there was a section on the Grand Canyon and how it was supposedly carved by floodwaters. I've written entire articles on this subject as well. Others have covered it extensively. Rather than rehash everything here, I'll just simply point out that the Grand Canyon, at a length of 300 miles, averaging one mile high and one mile wide, would have contained a lot of rock. Even before my recent work with AI, I was able to use my college math and physics to sketch out the approximate mass of rock that would have to be moved and the force of water it would have taken to move it.

The simple answer, without boring you with the math, is that the flood water would have had to be moving at such a speed and such force that it would have vaporized the rock, and done so roughly in a straight line. Yet, if you've ever seen photos of the Grand Canyon from space, or toured it in person, you know it consists of very long sections of oxbow curves, some turning back on themselves as much as 180 degrees.

How could a wall of water, moving at unimaginably high speed, suddenly slow itself down in time to take slow curves, then speed back up again, slow down... speed up... for 300 miles? My Geology 101 education from university came screaming to the forefront of my mind. Meandering rivers like the Colorado were formed by slow-moving water, over long time scales.

And why did this incredible gouge in the Earth only happen to a section of earth that was only around one mile wide? Did God have something against that particular patch of land and single it out?

To their credit, someone on our tour did remark, "Wow, that would be a lot of rock to move."

"Oh, it wasn't rock back then; it was dirt," somebody helpfully chimed in. "The rain from the flood changed it all to mud, so it was easy to carve it all out. It changed into rock right afterward. See all the rock?" the helpful scientist concluded.

My tongue was bleeding from all the biting. I worked on a pipeline crew to earn money for college. Ditch collapses were a big worry then, and now. Every year there's at least one news story of a ditch, maybe four to six feet deep, collapsing when the earth becomes wet and unstable. But our tour group nodded their heads in unison, in agreement, at the suggestion of two walls of mud, each a mile high, standing there long enough to turn to stone. Physically, it just isn't possible. I tried to flee the group again.

Oh Lord, help me… another gift shop.

I found the most disturbing part of the entire encounter on one of the upper decks. It was an exhibit that explained why all humans *deserve* to die. It took me back to my early church days, when nobody could sufficiently explain to me why God allowed all those innocent babies and children to drown, even if they did get to go to Heaven right after choking down all the water.

According to the Ark Encounter, we are all inherently evil, and the only thing that could save us was a human sacrifice (Jesus). Thankfully, rainbows meant that we would no longer have to worry about all being wiped out in a massive flood. But there was an even worse fate awaiting those who weren't "saved"—a one-way trip to Hell, where we would burn for all eternity. This was the true message of this "family attraction," and of my fundamentalist upbringing: an unyielding requirement to believe every word you're told, with a horrible judgment waiting for those who fell short.

Parents towing groups of young children walked by, the kids listening wide-eyed as all of this was explained to them. I felt sick, and fortunately found a bench to sit down. This was my childhood, being replayed in the theater of a multi-million-dollar boat that would have broken apart and sunk immediately if it ever saw water.

All of those bright young minds being wasted. All of that potential. Poisoned. How many were destined to spend their nights like I did, lying in bed crying, conflicted between a reality they could see with their own eyes and an artificial world where sloths, who travel at less than 0.2 miles per hour, somehow made the slow crawl to a magical boat before the hatch closed, amazingly avoiding being trampled by the dinosaurs?

I exited the ark, walking past another gift shop, the largest I'd seen, complete with food and treats for the kids. This place had to be raking in a small fortune.

I saw a sign that read "petting zoo" and headed down the path. There was a single goat, tied to a metal stake driven into the ground. Apparently, this part of the attraction was still under construction. I understand that a year or two after I left, they added zip lines, a merry-go-round, and other things that the Bible says were on Noah's Ark.

Separation of Church and State

In 2019, I'd begin a campaign to raise public awareness of what I viewed as a gross misappropriation of funds and the constitutionally-mandated separation of church and state. A public college, Southeast Kentucky Community Technical College (SKCTC), used scholarship money to take middle and high school students on a "field trip" to two Answers in Genesis "museums," which preached Young Earth Creationism. The so-called museums were undeniably not only religious in nature but proselytized heavily.

The field trip was heavily criticized by the American Civil Liberties Union, the National Center for Science Education, the Freedom from Religion Foundation, and prominent scientific scholars in Kentucky universities.

I somehow managed to land the only interview with SKCTC before their public relations department shut down all communication on the incident. I was writing for Patheos, a website that ostensibly balances both religious and non-religious viewpoints. Despite my credentials that included Forbes and other well-known publications, nobody other than Patheos seemed willing to touch a religious story.

It seemed clear that school children should not be subjected to proselytization. But Answers in Genesis (AIG), one of the sponsors, was inviting the controversy in hopes of, in the words of founder Ken Ham, ushering a case to the Supreme Court. Ham wanted to overturn the clear dictates of the constitution's Establishment Clause.

Ham publicly invited schools who'd received warnings from the Freedom from Religion Foundation, offering them free admission to the AIG attractions. He'd already received offers of *pro bono* legal support from an ultra-conservative law group, Alliance Defending Freedom (ADF), which promised to extend their free legal support to any school that would send students on these "field trips."

Before the college cut off my interview privileges, a representative denied that the outing was religious in nature, describing it as a "cultural enrichment experience." This was curious to me, as during my own visit to the Ark Encounter, literally every exhibit on the property mentioned God, or was religious, including a final exhibit that exhorted visitors, young and old alike, to accept Jesus so that they may be "saved."

I found nothing in this "attraction" to represent any culture other than fundamentalist evangelism.

Ken Ham openly bragged in an AIG publication about three students who'd visited his attractions deciding to convert to Christianity on the spot. His group's mission statement clearly stated that this type of outcome was the desired effect of their parks.

And yet, these students from impoverished southeastern Kentucky were allowed to travel to these places as part of what was, on the surface, a secular science program. Never mind that as these children traversed the levels of the Ark Encounter, they would encounter an exhibit that told them, point-blank, that as sinners, they deserved to die.

I'd heard this message decades before, when my church sent me to watch a movie where a whore drank blood and people burned in a lake of fire.

In my interview with the college spokesperson, this was all defended by the argument that nothing in the law prohibited students from encountering exhibits of a religious nature. Admittedly, this would be true. But proselytizing... telling school children that they were sinners and deserved death? And that they had only one, singular chance at salvation? This was not science or culture. This was not a school trip to admire an old Gothic church, or to gaze upon the amazing medieval religious artwork of renowned painters. This was the threat drilled into me from my first days in a fundamentalist evangelical church: believe or suffer for eternity. You earned the suffering. It was your fault. You were given a chance and you ignored it.

Science education? Hardly.

When presented with the evidence that the attractions on the field trip undeniably proselytized, my contact at the college pushed back, saying that the school's purpose was merely to "encourage free thought."

"How does one encourage free thought when presented with the alternative of eternal punishment?" I asked in a follow-up question.

Upon advice of legal counsel, the college decided to end our interview.

These incidents are common in the Bible Belt. In Kentucky, one county (Letcher County) used a loophole created by a fundamentalist Christian governor, allowing religious studies in schools. Curriculum was based on literature produced by a company whose sole business was to produce Sunday school lessons. Students were required to memorize Bible verses as part of their coursework. This same governor would later use an official government Twitter account to encourage Kentucky schoolkids to participate in "bring your Bible to school day."

It doesn't end there. I later participated in a Freedom of Information Act (FOIA) filing to obtain information on an evangelical Kentucky teacher who was allegedly teaching fundamental religious beliefs in his classroom. The case was not hard to believe: the teacher's profile, published on the school's state-run website, was filled almost entirely with his religious beliefs and his desire to lead his students to Jesus through his teachings. The distinction is important here: this was a school website, paid for by taxpayer funds. Separation of church and state? Not here.

Not surprisingly, my FOIA request had to go through an arbitrator who, wait for it... was the school district's superintendent, and was on record defending the teacher's proselytizing. The records I had requested had, to a computer science expert, been heavily scrubbed. When I asked for an explanation for the information the arbitrator admitted was missing, I was told that it contained information of a personal nature. The normal procedure here would have been to redact the PII (personally identifiable information), simply by removing or blacking out names and other info. The superintendent saw it differently, and decided to leave out *entire emails and documents*. How many? He refused to disclose that information.

My recourse under state law? Of course, it was to appeal to the arbitrator who had removed the information in the first place.

I could see, and can now clearly see, a world with secular education slowly slipping from our grasp. It no longer seems to be enough that Americans have the freedom to worship in any way they choose in their private lives. Now a particular set of those beliefs must be introduced into daily school and secular life.

To those who are delighted by this prospect and find no fault, I would simply ask: what if the religious teachings in your child's school were not of your own faith?

It does not take a vivid imagination to predict the reaction of the evangelical teacher who was the target of my FOIA request, had he been informed that he was suddenly required to teach Hinduism or Islam in his classroom.

Eclipse

Being an avid astronomer since childhood, it was only natural that I would fall into the hobby of chasing total eclipses of the sun across the planet. Total solar eclipses occur when the moon, being approximately 400 times smaller than the sun, perfectly aligns with and blocks the sun, which is about 400 times farther away. We get about two chances per year to see this event, but the geometry and physics often conspire against us.

The moon's jaunt around the orbital plane of our planet is at an approximate five-degree angle, meaning our companion often passes above or below where it would cross the face of the sun. Also, the moon's orbit isn't circular. It's an ellipse. So, sometimes it is closer to the planet, but sometimes farther away—too far away to completely cover the sun.

When the moon does cross the face of the sun with an apparent size large enough to completely block our star, the size of the shadow it casts on the planet is so small that for any given place on the Earth, the chances of seeing a total eclipse are around once every 350 years for a given location. So, eclipse chasers travel the globe to put themselves in the right spot to collect "shadow time."

To add to the improbability and wonder, our planet is slowly losing its moon. Due to a principle known by the fancy term "conservation of angular momentum," our natural satellite is receding from us at a rate of about 1.5 inches per year. Simply put, the moon tugs on our planet, creating a "tidal bulge" that is slowing the speed of Earth's rotation. The lost momentum must, by the laws of physics, go somewhere. And it does—it's transferred to the moon. This isn't a wild conspiracy theory dreamed up by drug-influenced astronomers sitting in a cold mountaintop observatory on a long winter's night.

The Apollo astronauts left behind mirrors on the moon when they departed. By hitting those mirrors with lasers and measuring the time it takes to make the round trip, astronomers know for certain this is really happening.

I always joke to my friends, who I desperately try to convince to experience the joy of a total solar eclipse, "You better hurry, the moon is leaving us." In truth, the rate at which we're losing our neighbor isn't of great concern. In the billions of years it will take to make enough of a difference, we'll face much greater problems. If climate change doesn't do in our planet in the future, we're faced with the hard fact that our sun, like all stars, has a limited amount of fuel. At some point, around five billion years from now, the sun's mass will decrease from burning through so much of its fuel supply, and gravity won't be enough to contain the leftover hydrogen. The sun will begin to "puff out" and swallow the inner planets, including Mercury, Venus, and possibly the Earth.

To me, this makes this an even more amazing time to be alive. It's hard to describe the wonder of an eclipse in words. The sun disappears during what was just previously broad daylight. The stars and planets come out. A false sunset circles the horizon in a 360-degree swath. I've seen temperatures drop twenty degrees in a matter of a few minutes. Animals get confused. Birds flock to trees, thinking night has fallen. If you're close to a farm, you can see animals start their habitual sunset walk to the barn. Crickets begin chirping. Dew begins to form. For a maximum of approximately seven minutes and thirty seconds (though, sadly, often much shorter), night falls.

The sun's corona—it's atmosphere—shines as a milky white cloud around what looks like a black hole in the sky. Depending on magnetic conditions deep within the sun, the corona can be a simple, smooth, white cloud extending outward along the plane of the sun, or it can be an irregular, spikey cloud that has reminded me of a flower whose petals somehow evolved badly. Along the darkened silhouette of the moon, small white dots—Bailey's Beads—sometimes appear. These are rays of sunlight shining through mountain passes along the edge of the moon.

Eclipses are magical. The only other natural event I've ever seen that I can compare them with are the births of my nieces and nephews. In both situations, you just stand there and stare, asking yourself how this could be possible.

For all the years I've been in the hobby, I so looked forward to the year 2017, when a total eclipse would cross Kentucky. I'd been telling eclipse stories to my family for decades. This was just something they *had* to see. I rented cabins in the western part of the state, in the shadow's path, and begged, pleaded, and cajoled my loved ones to make the four-hour drive to see the event.

Clouds are the biggest enemy of eclipse chasers. I fretted over weather reports and predictions for weeks leading up to the event. It was during Kentucky's summer thunderstorm season, when clear blue skies could turn into tropical downpours at the drop of a hat.

Fortunately, on "E-Day," the skies were crystal clear at dawn, and remained that way. We were actually going to see this one!

My family, full of gifted photographers, lined up their tripods and cameras on the lawn outside our cabin. We watched, through solar filters I'd made for the telephoto lenses, as "first contact" occurred—the moon took its first bite out of the sun.

As the eclipse progressed, my young science-loving nieces shouted in joy. They'd set up an experiment, a simple white sheet laid out underneath a large oak tree. I'd told them that the leaves of the tree could act to form tiny pinhole cameras and project thousands of images of the sun across the ground. They pointed proudly to hundreds of tiny images of white discs being swallowed up by black discs on the sheet. Kids truly do get excited about science. You just have to make the effort to involve them.

The big moment finally came: a "diamond ring," the last bit of sunlight escaping along the edge of the moon, looking like a huge diamond shining in the sky. And then, darkness. Except for a black hole in the sky where the sun once stood, surrounded by a glowing white corona.

There were some cheers, and then silence. Except for the occasional "oh my God." As our eyes adjusted to the darkness, stars and planets began to appear in the sky. For most of my family, this would be their first-ever glimpse of the planet Mercury, which orbits so close to the sun that it's normally lost in glare at sunrise and sunset. Venus, the evening/morning planet, was nearby. I made sure to point these out, but otherwise remained silent, allowing my family to form their own impressions.

The sounds of camera shutters took over as everyone finally remembered to trigger their cable releases. There are many stories of first-time eclipse watchers who are so enthralled that they forget to take a picture. To be honest, photos aren't necessary. Many seasoned chasers don't even bother with cameras. They're just a distraction. These are events to be experienced and cherished. Photos are nice to have, but absolutely nothing can recreate the feeling of being there.

The end of totality was announced, as always, by a second diamond ring. Cheers went up, and hugs went around. I don't think I was the only one with a tear in my eye.

I walked over to my father, a devout Christian raised in the same fundamentalist faith of my youth, and asked, "Well, what did you think?"

"I don't know how anyone could ever see this and not believe in God," he said.

A chill went down my spine. These were *exactly* the words I'd uttered to my cousin decades before, on a night under the Kentucky skies when we watched a lunar eclipse together.

The exact words, to the letter.

My perspective on the heavens and natural causes had changed a lot since that night with my cousin, but as I looked at my father, I understood he was feeling all the wonder of nature that I had felt, and that I still feel. A deeply religious man, he was expressing himself in the same way I had done, and there was an understanding between us.

I wasn't going to go into scientist mode on this day. To have done so would have been cruel. It would have taken away some of the wonder. I could have blathered on about orbits and geometry and apparent sizes and religion having nothing to do with it. But I'd invited my loved ones to this spot to share the wonder of the most incredible event in nature, and it was obviously having an impact. Who was I to take away their joy?

I put my arm around Dad's shoulder and simply said, "Yeah, that was amazing."

There are times when it's best just to be quiet and let people enjoy things in their own way, according to their own understanding. This was one of those times. Dad wasn't trying to force a religious belief on anyone. He was simply expressing his feelings. As a non-believer, I confess there are times when I must force myself to step back just a little and not automatically go into full-science mode.

Sometimes you hug the loved one in your life, wipe away the tears in your eyes as you realize you've both experienced the wonders of the universe—albeit from a different perspective—and just enjoy the moment.

Dad was now hooked on the wonder of solar eclipses. I mentioned that we would have a chance to witness this rare event in just seven years, in April 2024, when the moon's shadow would once again cross Kentucky. My father and I started making plans.

I could not have known at the time that cancer was growing in his body, and our time together would soon come to an end.

Shadow of Death

My father was diagnosed with prostate cancer many years before, but doctors concluded that the disease was caught very early and stressed that we should not be overly concerned. Treatment with radiation led to the eventual pronouncement that he was fully in remission, and we put the matter behind us.

Until the cancer returned.

The discovery was a surprise to everyone, including Dad's doctors. It wasn't just his prostate that was affected now. The disease had spread to surrounding organs and had reached as far as his kidneys. Cancer experts tried everything—chemotherapy, new experimental immune system treatments that had shown promise—but nothing worked.

The illness was very aggressive and in a very short time, Dad was in palliative care (during Covid, which added to our family's pain as we could not be with him). Then, finally, came the phone call.

My father had passed away.

Our promise we made during the 2017 solar eclipse, to stand together again one day in the moon's shadow, would never be fulfilled. There was a new shadow to stand in now. The shadow of death.

As a non-believer, it was not possible for me to find any relief or hope in religion. Honestly, especially in the immediate aftermath of his passing, I'm not sure even our most religious family members found much comfort in their faith. The immediate shock and feelings of losing a loved one are just too raw. I think this is why even the most devout Christians cry so hard at funerals. Even though they believe they'll eventually be reunited, the sense of loss is just too palpable. We're only human, after all.

People often ask me how I deal with a loss like this, given that I have no belief in an afterlife. This was not the first tragic loss in our family. I'd lost a sister to the same disease several years before.

My answer is that I hold onto the memories that I have of my loved ones. I have enough precious remembrances of my dad to fill several books.

The most visceral comes from a Milwaukee snowstorm when I was three years old. It was nighttime. Dad bundled me in my snowsuit and carried me out to my Flexible Flyer sled, and we set off into the snow.

Milwaukee snowstorms, with their lake effect snow, are a wonder to behold. As a toddler, the fluffy white stuff would soon accumulate to a height greater than mine, giving the sense of traveling through a long white tunnel as Dad pulled me down the sidewalk.

Today, almost sixty years later, I can still see and feel the moment as if it just happened yesterday. The incredible muffled silence that comes with heavy snowfall… nothing but the sound of the sled runners on the snow and that amazing (no pun intended) white noise produced by countless flakes landing around me. Laying on my back on the sled and looking up at the sky, huge snowflakes drifting down like falling stars, landing on my cheeks. Sticking out my tongue and trying to catch them as they fell.

But most of all, the memory of my father. The always strong, loving man ahead of me, pulling me through our neighborhood, a big smile on his face as he reveled in his young son experiencing one of the small wonders of life. He laughed as I chased the snowflakes with my tongue.

Time to go inside? No. "One more time daddy, please. Just one more time around the block."

Of course, he said yes. He altered our route, found a small hill, and let me gleefully ride down. Over and over, until the Wisconsin cold finally determined it was time to head home.

Back inside, there was Mom, standing at the stove, holding my first-born baby sister, making hot chocolate. Not today's cheap packeted version of the drink. Authentic Hershey's Cocoa mixed with liberal amounts of sugar and whole milk—none of the lame 2% stuff. Real hot chocolate, with marshmallows. We sat at the kitchen table together and drank something that the far-in-the-future Starbucks would never be able to equal.

My little sister giggled and ate marshmallows.

I looked across the table at my family with love. This was a magical evening. It was burned not only into my memory, but if I may wax poetic, it was burned into my heart. When I want to be with my father now, in those moments when I miss him so deeply, I close my eyes and return to the snowstorm.

There are countless similar memories. Not just of Dad, but all the loved ones I've lost. And for the family members I'm still blessed to have, I continue to burn into my mind our experiences together. Not believing in an afterlife, I feel the events occurring daily in my life, the memories I'm building now, may possibly mean more to me than if I was religious. I mean this in a positive way—every moment I spend with a friend or family member now, I keep in mind it may be our last moment. I value the time with them even more.

This is how I deal with the loss of a loved one. When I say I feel my father with me when I close my eyes, I don't mean it in a spiritual or visceral sense, as if he's truly in the room with me. But I do feel him. I see him.

My understanding of reality does not provide me with the comfort of believing I'll actually physically stand next to this great man again.

However, I tell my religious friends that, in all sincerity, if there's anything in the world I'm wrong about, I hope it would be this.

Conclusion

Atheists are often criticized by believers for allegedly having no purpose in life. I find this odd, as I can go back to the earliest days of my existence, to an age when some scientists debate whether I would have been able to form true memories at all, and I can clearly remember an overriding sense of purpose:

I wanted to be a good person.

Early on, I looked solely to my parents, and then to my immediate family, for confirmation that I was accomplishing this goal. Later, as I began my journey into religion, I began to look to God for that same affirmation.

By all accounts, I was achieving my goal. My parents and family were proud of me. Teachers praised not just my intellect but my behavior in class. Strangers would compliment my parents on the fine son they were raising. Employers commended my work ethic and told me they wished they had more employees like me. But most of all, my church loved me, which meant God loved me. I was devout; I was evangelical about spreading the religion, and I could quote scripture as well as any preacher.

All of that changed, seemingly overnight, when I "came out" as non-religious. On social media, that great indicator of how important we really are in the world (sarcasm intended), I saw the first indicator that something was wrong. I lost nearly fifty percent of my friends. This was followed by shunning from friends and family members, even to the point of being disowned.

I was still the same person. My morals hadn't changed. None of the things that people had praised me for, during my entire life, were gone: except the religion. I'd even been acting the part of a good Christian for quite some time, afraid to tell people how I really felt; so even as a non-believer, I still met the religious criteria for being a godly man. But once the truth came out, I was immediately lumped into the "ugly atheist" group along with Madalyn O'Hair and all the other "evil" people my church had demonized when I was young.

Despite all of this, my main purpose in life has not changed one iota. I still want to be the best person I can be. My motivations have shifted somewhat, however. I no longer have the threat of eternal punishment hanging over my head, so I don't perform good acts out of fear. The promise of an eternity in the paradise of Heaven is gone, so when I feed a homeless person, I don't expect a reward. My focus on who cares whether I'm a good person is more personal. The first person I look to for approval of my actions is myself. Am I proud of the way I am behaving? Do I loathe myself for being a failed person, constantly seeking heavenly approval, as fundamentalism had taught me, or did I love myself as a normal human being doing his best to make the world a better place?

Given this newfound confidence in myself, accepting responsibility for my own actions and working to be the best person I can be, how do those I care about view me? Am I seen as a trustworthy person? Are my relationships with my loved ones stable? Do I have the love and respect of my family and friends? When I make a promise to, say, my employer, do they trust my integrity and know that I will get the job done?

Am I "paying forward" all the time and hard work that my teachers and mentors invested in me? Do I do enough to help the next generation gain a solid education, so that they might have a better chance in life? These are the questions that are important to me now.

I'd like to be able to say that I steer completely clear of discussions of religion, but that would be a lie. Throughout my life, especially in the last few decades, I've noticed a very disturbing shift in my country—the embracing of many of my old fundamentalist beliefs by a growing segment of our society. Conservative religious principles are invading our schools and government.

Science rejection is rampant. Books are being removed from libraries, denying not just an offended parent's child the chance to read, but removing the literature from the reach of every other student in the school. A small but vocal group of fundamentalists, deciding on behalf of everyone else what can or can't be read; this is one of the many things I feel obliged to speak out against.

The school prayers that were ruled invalid during my school years have returned with the backing of the Supreme Court. I can speak from personal experience how uncomfortable it would be as a non-believer not to take a knee to pray at midfield in front of a crowd, as my coach led a Christian prayer. The peer pressure is enormous. I certainly support the right to freedom of religion but not a public employee forcing others to pray while he/she is on the clock, supported by the taxpayer's dime.

As I mentioned earlier in the book, fundamentalist organizations are working hard to reintroduce Bible study into public classrooms. Kentucky's former governor (Matt Bevin) sponsored a "bring your Bible to school day." I researched and wrote an article about a Kentucky college taking high school students on "field trips" to the Young Earth Creationist Ark Encounter. I've discovered public schools using Bible study courses as their curriculum. Despite a Supreme Court ruling against prayer in schools, handed down before I was ever born, fundamentalist religion has never truly left Kentucky schools (or, apparently, other schools in the Bible Belt).

Topics such as evolution, that would have been of great benefit to me as a young student in the field of science, are simply not taught in many schools. Evangelical parents protest too loudly when attempts are made. Likewise, important discussions about seemingly anything and everything related to human sexuality are banned. I still recall with sadness my high school friend who became pregnant before her sixteenth birthday, and how easily basic education could have prevented that.

Fundamentalist religious beliefs have crept their way into our government. From seemingly trivial laws such as not being allowed to buy alcohol on Sunday in Kentucky until church services are over, to much more serious problems such as having faith-based laws regulating reproductive rights. We now have discriminatory laws affecting persons of particular sexual orientations. Faith seems to have become a driving factor in determining our laws. A small but very vocal group of high-ranking politicians are openly declaring that we are a Christian nation, and routinely work to introduce non-secular legislation in our House and Senate.

I see cause for worry here. I grew up in a very exciting time in the United States. I watched the first moon landings with my father. Advances in medicine saved my mother from leukemia. The advent of computer technology has accelerated us into a new age—The Information Age. As a species, we have so much potential. And yet, as I lived through all these advances, I had the constant backward pull of fundamentalism, denying all the progress I was experiencing.

I feel that we, as a country, are being dragged back in time.

Even as I reached my personal conclusion that there was no

evidence of a supernatural creator, or a being that guided our lives, I was still able to maintain meaningful relationships with those who had such beliefs. Indeed, I learned many positive life lessons from those who were believers. The difference—the very important difference—is that these important influencers did not try to force their beliefs on me.

I'd be more than happy to step back and completely disengage from any debate regarding religion, but the fact is that a certain group, fundamentalist believers, won't allow that to happen. Great damage was done to my emotional health, as well as my educational development, by the same crowd now banning books and passing faith-based laws. I can't stand back and watch. I can't remain quiet.

My deepest regret is that I am relatively unknown in this world. I am not famous. I do not have a large public platform from which to speak. I have only my writing, which, given that there's an average of 7,500 books published daily on Amazon.com, statistically does not have a great chance of reaching a significant amount of people.

And yet, I must speak out.

Having suffered through indoctrination as a child, I wanted to weep for the children I saw boarding Answers in Genesis' Ark Encounter upon my visit there. How many of those kids are going to go through all the years of nightmares and conflict that I suffered? Our country has consistently been sliding down the global ranks of most-educated nations. As we continue to remove science from our schools, how much further will we fall?

My journey to reason has placed me into a group that current polling shows is among the most mistrusted in the United States, even though other polls show my fellow non-believers and I are among the most religiously literate. Despite this, I have not deviated from my belief that I am being the best person I can be and am not discouraged by continuing my attempts to be even better.

Part of that effort is to do all I can to warn others of the dangers we are now presented with via religious fundamentalism. I hope you will join me in that effort.

One Last Thing

If you enjoyed this book, please consider leaving a rating and a short review on Amazon. Independent writers live (or die) based on reviews. With an estimated 7,000 books published each day, every review, every star, every kind word… they mean so much to authors. It is hard to stand out in such a crowded marketplace. Your kind consideration would be greatly appreciated.

About the Author

Mark Alsip has a bachelor's degree in computer science and a minor in mathematics, with a heavy concentration in natural sciences. The latter led to an additional two-year stint as a pre-med student, where he fell in love with biology. He was a member of the Lockheed IDEX II team, a recently declassified aerospace project now enshrined in the Smithsonian's National Air and Space Museum. With forty years of Information Technology experience, his favorite projects by far have involved space, medicine, and the natural sciences.

His childhood indoctrination into a fundamentalist, science-denying religious sect, and the inevitable conflicts with science, history, and reason have played an important role in shaping his life. While not anti-religious, his experiences have led him to believe that extreme religious fundamentalism is an existential threat to society and democracy.

He began writing the popular skeptical science blog "Bad Science Debunked" in the early 2010s, taking on scientific misinformation found on social media, and, his favorite topic, doctors and new-age health "experts" who used scare tactics about "chemicals in our food" to sell their own alternative health supplements, which, unsurprisingly, contained the same ingredients they labeled "toxic." And yes, by the way, all foods are completely and entirely made up of chemicals.

Mark's skeptical writing has been featured in publications such as Forbes, Skeptical Inquirer, and the newsletter of the National Center for Science Education. He has been a frequent op-ed columnist for Kentucky newspapers, where he continues his mission to debunk science denial and misinformation.

"Journey to Reason" is Mark's first solo book, but he greatly enjoyed being a co-author on a previous tome, "The Fear Babe," where he expanded upon the theme of his science blog, taking on the organic food industry, which has seemingly failed to realize that 100% of foods are made of chemicals, and that crude oil and cyanide are "natural and organic"—but you wouldn't want to consume them.

In his spare time, Mark enjoys astronomy, amateur radio, photography, and especially, traveling the world with his wife, Melissa. Together, they have had the privilege of visiting over forty countries. He is a die-hard chaser of rare total eclipses of the sun, and, if the weather is clear in Texas in April 2024, he will have bagged his tenth such event. Few things in nature compare to the beauty of the sun being extinguished in the middle of the day, with the stars and planets coming out.

This is an amazing world we live in, and the author strives to enjoy as much of it as possible.

Acknowledgements

To all the teachers and mentors, the unsung heroes who instilled in me a lifelong joy for learning: Thank you. I hope I can pay forward what you've done for me.

To my parents, you are wonderful beyond words. Even when you didn't understand my crazy love for science and technology, you encouraged me, and stood behind me every step of the way.

Michael Pilgrim, without your amazing editing skills, this book would not have been possible. I am grateful you were willing to take on a new writer full of ideas but so lacking in knowledge of how the editing process works.

Melissa Alsip, your proofreading added the final layer of polish this book needed.

Alexander von Ness, your cover designs are works of art. I am proud to have my name on one of your covers.

Blackbox Media, your work turned my promotional efforts from so-so to something absolutely stellar!

Saving the most important for last, to my dear wife Melissa: Thank you for understanding me, as well as pretending to understand me when maybe you really don't. I'm not sure how many other wives could feign interest when their husband wakes them up in the middle of the night to explain a fatal flaw he found in a Young Earth Creationist article on radiometric dating. Your love for me, and your belief in me... those are the things I fall back on when I struggle to keep moving forward.